Advanced Mathematics-II

Advanced Mathematics-II

[Bridge Course for IV Semester Engineering Students
of VTU with Polytechnic Background]

Chidanand S. Mujawar

Professor of Mathematics
S.G. Balekundri Institute of Technology, Belgaum
KLE's College of Engineering and Technology, Belgaum
Former Head, Dept. of Mathematics
&
Visiting Professor
P.G. Department, V.T.U., Belgaum

I.K. International Publishing House Pvt. Ltd.

NEW DELHI • BANGALORE

Published by
I.K. International Publishing House Pvt. Ltd.
S-25, Green Park Extension
Uphaar Cinema Market
New Delhi–110 016 (India)
E-mail: info@ikinternational.com
Website: www.ikbooks.com

ISBN 978-93-81141-87-8

Published by Krishan Makhijani for I.K. International Publishing House Pvt. Ltd. S-25, Green Park Extension, Uphaar Cinema Market, New Delhi–110 016. Printed by Rekha Printers Pvt. Ltd., Okhla Industrial Area, Phase II, New Delhi–110 020.

Dedicated
to
His Holiness
Shri Siddharam Swamiji
Rudrakshimath, Naganur, Belgaum

"The man who makes no mistakes does not usually make anything."

— **Bishop W.C. Magee**

"If the facts don't fit the theory, change the facts.

—**Albert Einstein**

Foreword

Advanced Mathematics-II will serve as a good textbook for bridge course of the fourth semester engineering students of Visvesvaraya Technological University with polytechnic background. The presentation is simple, rigorous and compatible with the curriculum of Engineering Mathematics.

It gives me immense pleasure to introduce the book Advanced Mathematics-II, by Prof. Chidanand S. Mujawar the publication of which heads the completion of book that caters completely and effectively from a modern point of view of engineering students. This book has been organized and executed with lot of care, dedication and passion for lucidity. The chapters are largely independent covering Analytical Solid Geometry, Vector Algebra and Calculus and Laplace Transforms.

The author has been an outstanding teacher for over 38 years. He has taught for 28 years at K.L.E. Society's College of Engineering and Technology affiliated to VTU, Belgaum. He deserves our praise and encouragement for accomplishing this task. I wish the users good luck and happy reading.

Dr. H.G. Shekharappa
Co-ordinator Postgraduate Centre
Visvesvaraya Technological University
Belgaum

Preface

Advanced Mathematics-II is designed for the bridge course of engineering students of Visvesvaraya Technological University with polytechnic backgound. It is intended as a text for undergraduate students of engineering with polytechnic background. The book mainly contains Analytical Solid Geometry, Vector Algebra and Vector Calculus and Laplace Transforms. Each topic is treated in a systematic and logical manner. An outstanding and distinguishing feature of the book is the large number of typical solved examples followed by well-graded question bank for each topic. Many examples and problems have been selected from recent papers of various engineering examinations.

I take this opportunity to thank my colleagues for suggestions for the improvement of this book.

I will gratefully accept from the readers any corrections, criticisms and suggestions for the improvement of the book.

Chidanand S. Mujawar

Contents

PART (A)
ANALYTICAL SOLID GEOMETRY

Syllabus for Bridge Course
(Common to all Branches)

(A) ANALYTICAL SOLID GEOMETRY

UNIT 1

Distance formulae (without proof), Division formula, Direction cosines and Direction ratios.

Unit 2

Planes and straight lines, Angle between straight lines and planes.

(B) VECTOR ALGEBRA AND VECTOR CALCULUS

Unit 3

Vector addition, multiplication, Dot and cross products, Triple products and problems.

Unit 4

Vector differentiation, Velocity, Acceleration of a vector point function, gradient, divergence and curl, solenoidal and irrotational vector fields, simple and direct problems.

(C) LAPLACE TRANSFORMS

Unit 5

Definition, Laplace transform of elementary functions, Laplace transform of derivatives and integrals, Multiplication by t^n and division by t and examples.

Unit 6

Inverse Laplace transforms, Applications; Laplace transforms of linear differential equations and simultaneous differential equations, simple and direct problems.

PART (A)
Analytical Solid Geometry

Analytical Solid Geometry

Let the three mutually perpendicular lines $X'OX$, $Y'OY$ and $Z'OZ$ intersect at O. Then 'O' is called the origin. $X'OX$ is called the x-axis, $Y'OY$ the y-axis and $Z'OZ$ the z-axis and taken together these are called the co-ordinate axes. The plane YOZ is called the YZ-plane, the plane ZOX the ZX-plane and the plane XOY the XY-plane and taken together are called the co-ordinate planes.

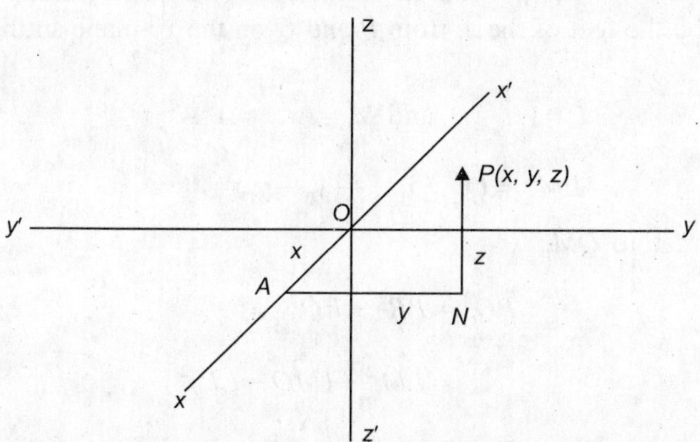

To find the co-ordinates of a point P, draw to $PN \perp$ to XY-plane. From N draw $NA \perp$ to the x-axis. $OA = x$, $AN = y$, $NP = z$, then $(x,. y, z)$ are the co-ordinates of P.

At N, $z = o$, therefore, the co-ordinates of N are $(x, y, 0)$.

Similarly, at L, $x = 0$ the co-ordinates of L are $(0, y, z)$ and

at M, $y = 0$, the co-ordinates of M are $(x, 0, z)$.

The co-ordinates of the points on the axes are $(x, 0, 0)$, $(0, y, 0)$ and $(0, 0, z)$ respectively. The co-ordinates of the origin O are $(0, 0, 0)$. The x, y, z co-ordinates are positive along OX, OY and OZ respectively and negative along OX', OY' and OZ' respectively.

The three co-ordinate planes divide the space into eight compartments called octants. The octant $OXYZ$ in which all the co-ordinates are positive is called positive or the first octant.

DISTANCE BETWEEN TWO POINTS

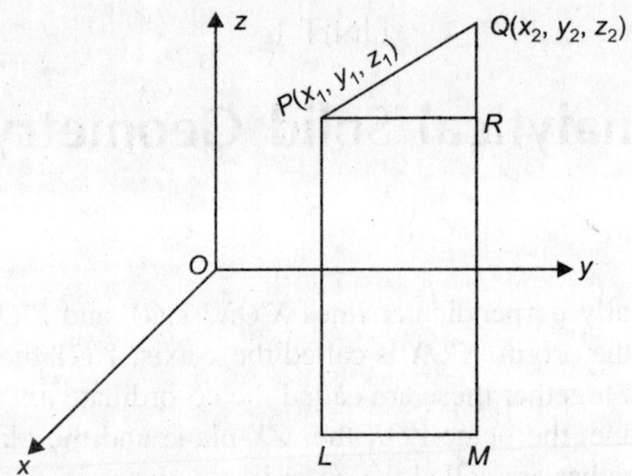

Let (x_1, y_1, z_1) and (x_2, y_2, z_2) be the co-ordinates of two points joining the line PQ. Let L and M be the feet of the $\perp$ from P and Q on the xy-plane so that in the xy-plane.

$$L = (x_1, y_1) \text{ and } M = (x_2, y_2)$$

$\therefore \qquad LM^2 = (x_2 - x_1)^2 + (y_2 - y_1)^2$

Now, draw $PR \perp$ to QM

$\therefore \qquad PQ^2 = PR^2 + RQ^2$

$$= LM^2 + (MQ - LP)^2$$

$$= (x_2 - x_1)^2 + (y_2 - y_1)^2 + (z_2 - z_1)^2$$

$\therefore \quad PQ = \sqrt{(x_2 - x_1)^2 + (y_2 - y_1)^2 + (z_2 - z_1)^2}$

Cor: Distance of the point $P(x, y, z)$ from the origin $O(0, 0, 0)$ is given by

$$OP = \sqrt{x^2 + y^2 + z^2}$$

SECTION FORMULAE

Let $R(x, y, z)$ divide PQ in the ratio $m_1 : m_2$ where $P \equiv (x_1, y_1, z_1)$ and $Q \equiv (x_2, y_2, z_2)$. Draw PL, QM, RN $\perp$s to the xy-plane. Through R draw SRT parallel to LM meeting PL (produced) in S and QM in T.

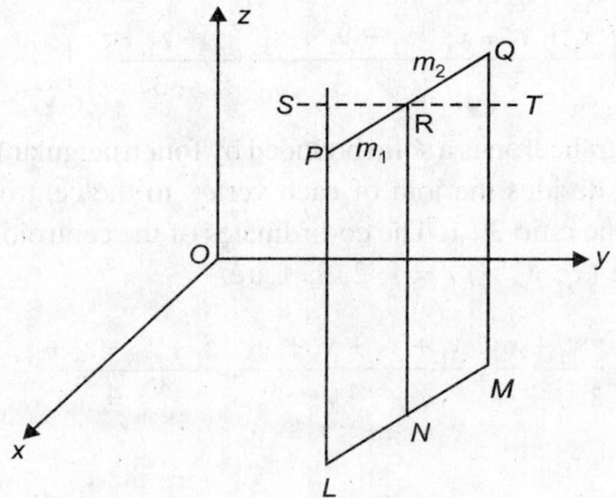

Then from the similar Δs *PSR* and *RTQ*, we have

$$\frac{m_1}{m_2} = \frac{PR}{RQ} = \frac{PS}{TQ} = \frac{LS - LP}{MQ - MT}$$

$$= \frac{NR - LP}{MQ - NR} = \frac{z - z_1}{z_2 - z}$$

$\therefore \qquad\qquad m_1 z_2 - m_1 z = m_2 z - m_2 z_1$

$\therefore \qquad\qquad z = \dfrac{m_1 z_2 + m_2 z_1}{m_1 + m_2}$

Similarly, we can prove, $x = \dfrac{m_1 x_2 + m_2 x_1}{m_1 + m_2}$ and $y = \dfrac{m_1 y_2 + m_2 y_1}{m_1 + m_2}$

$\therefore$ The co-ordinates of *R* are

$$\left(\frac{m_1 x_2 + m_2 x_1}{m_1 + m_2}, \frac{m_1 y_2 + m_2 y_1}{m_1 + m_2}, \frac{m_1 z_2 + m_2 z_1}{m_1 + m_2} \right)$$

Cor 1: Midpoint formula

If *R* divides *PQ* at the midpoint then $m_1 = m_2$,

$\therefore$ The co-ordinates of the midpoint are

$$\left(\frac{x_1 + x_2}{2}, \frac{y_1 + y_2}{2}, \frac{z_1 + z_2}{2} \right)$$

Cor 2: The co-ordinates of the centroid of the triangle whose vertices are (x_1, y_1, z_1); (x_2, y_2, z_2) and (x_3, y_3, z_3) are given by

$$G\left(\frac{x_1 + x_2 + x_3}{3}, \frac{y_1 + y_2 + y_3}{3}, \frac{z_1 + z_2 + z_3}{3}\right)$$

Tetrahedron: A tetrahedron is a solid bounded by four triangular faces. The centroid of the tetrahedron divides the join of each vertex to the centroid of the opposite triangular face in the ratio 3 : 1. The co-ordinates of the centroid of the tetrahedron whose vertices are (x_r, y_r, z_r) $r = 1, 2, 3, 4$ are

$$\left[\frac{x_1 + x_2 + x_3 + x_4}{4}, \frac{y_1 + y_2 + y_3 + y_4}{4}, \frac{z_1 + z_2 + z_3 + z_4}{4}\right]$$

Examples

(1) Show that the points $P(3, 2, -4)$, $Q(5, 4, -6)$ and $R(9, 8, -10)$ are collinear.

Solution: Let the point $Q(5, 4, -6)$ divide the line PR in the ratio $m : 1$

$\therefore$ The co-ordinates of Q are

$$\left(\frac{9m+3}{m+1}, \frac{8m+2}{m+1}, \frac{-10m-4}{m+1}\right)$$

But the co-ordinates of Q are given to be $(5, 4, -6)$

$$\therefore \qquad \frac{9m+3}{m+1} = 5 \text{ or } 9m+3 = 5(m+1) \;\therefore\; 4m = 2 \;\therefore\; m = \frac{1}{2}$$

Hence, Q divides PR internally in the ratio $\frac{1}{2}:1$ or $1 : 2$

$\therefore$ the points P, Q, R are collinear.

(2) Show that the points $(0, 4, 1)$, $2, 3 -1)$, $(4, 5, 0)$ and $(2, 6, 2)$ are the vertices of a square.

Solution: Let the given points be $A(0, 4, 1)$, $B(2, 3, -1)$ $C(4, 5, 0)$ and $D(2, 6, 2)$.

By distance formula

$$AB = \sqrt{(2-0)^2 + (3-4)^2 + (-1-1)^2} = \sqrt{4+1+4} = 3$$

$$BC = \sqrt{(4-2)^2 + (5-3)^2 + (0+1)^2} = \sqrt{4+4+1} = 3$$

$$CD = \sqrt{(2-4)^2 + (6-5)^2 + (2-0)^2} = \sqrt{4+1+4} = 3$$

$$DA = \sqrt{(0-2)^2 + (4-6)^2 + (1-2)^2} = \sqrt{4+4+1} = 3$$

$$\therefore \quad AB = BC = CD = DA$$

$\therefore$ *ABCD* is either a square or rhombus.

(3) Show that the points $(-4, 9, 6)$, $(-1, 6, 6)$, $(0, 7, 10)$ form a right-angled isosceles triangle.

Solution: Let $A(-4, 9, 6)$, $B(-1, 6, 6)$ and $C(0,.\ 7, 10)$ be the vertices of a triangle.

$$AB = \sqrt{(-1+4)^2 + (6-9)^2 + (6-6)^2} = \sqrt{9+9+0} = \sqrt{18}$$

$$BC = \sqrt{(0+1)^2 + (7-6)^2 + (10-6)^2} = \sqrt{1+1+16} = \sqrt{18}$$

$$CA = \sqrt{(0+4)^2 + (9-7)^2 + (10-6)^2} = \sqrt{16+4+16} = \sqrt{36}$$

$\therefore \ AB^2 + BC^2 = AC^2$

Hence, the points *A, B, C* form an isosceles triangle.

ANGLE BETWEEN TWO NON-INTERSECTING LINES

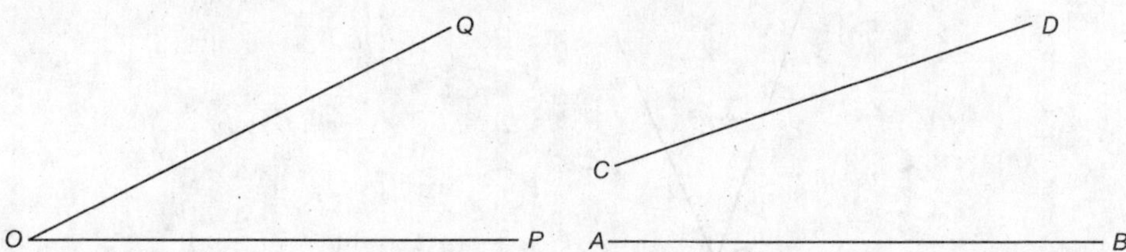

Angle between two non-intersecting lines is the angle between their parallels through any point. Thus, the angle between the lines *AB* and *CD* = $\angle POQ$ between their parallels through *O*.

Direction Cosines

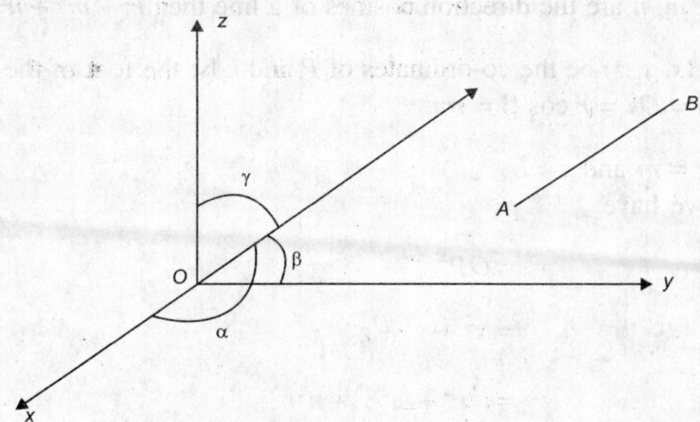

If any line *AB* makes angles α, β, γ with the positive direction of the axes then cos α, cos β, cos γ are called the direction cosines of the line *AB*, which are usually denoted by, *l, m, n*.

i.e., $\qquad\qquad l = \cos\alpha, \; m = \cos\beta, \; n = \cos\gamma$

Cor 1: Direction cosines of the axes.

As *x*-axis makes angles $0, \dfrac{\pi}{2}, \dfrac{\pi}{2}$ with the axes.

$$\therefore \qquad \cos 0 = 1, \; \cos\frac{\pi}{2} = 0 \text{ and } \cos\frac{\pi}{2} = 0$$

∴ The direction cosines of *x*-axis are (1, 0, 0)
 Similarly, the direction cosines of *y* and *z* are (0, 1, 0) and (0, 0, 1) respectively.

Cor 2: If *l, m, n* are the direction cosines of a line *OP*(= *r*) then the co-ordinates of the point *P* are (*lr, mr, nr*).

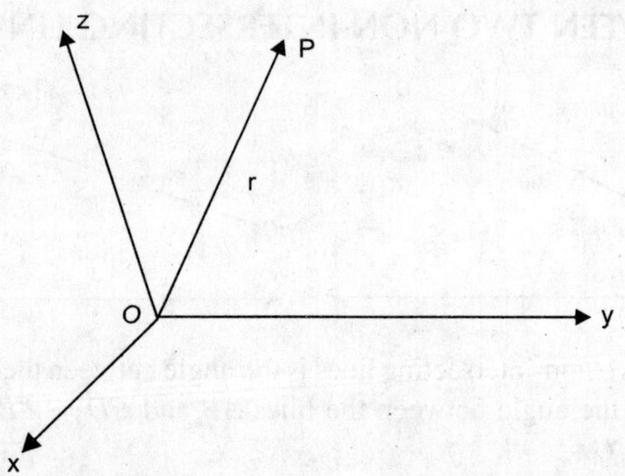

Cor 3: If *l, m, n* are the direction cosines of a line then $l^2 + m^2 + n^2 = 1$

Proof: Let (*x, y, z*) be the co-ordinates of *P* and *k* be the foot of the ⊥ from *P* on *Oy*. Then $y = Ok = r\cos\beta = rm$.

Similarly, $z = nr$ and $x = lr$
 Now, we have

$$r^2 = OP^2$$

$$= x^2 + y^2 + z^2$$

$$= l^2 r^2 + m^2 r^2 + n^2 r^2 \qquad\qquad \text{by Cor 2}$$

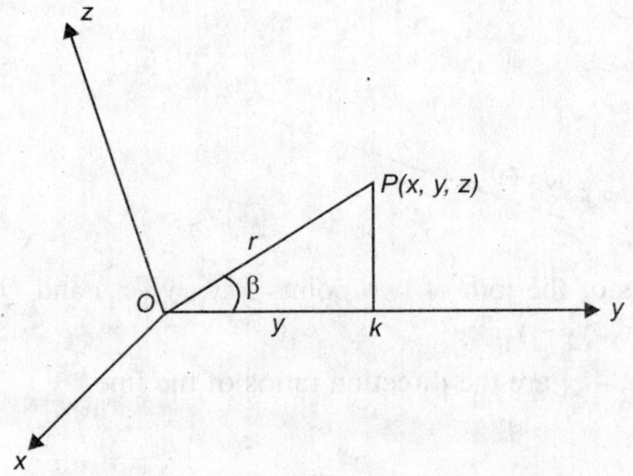

$$r^2 = r^2(l^2 + m^2 + n^2)$$

$\therefore$ $\boxed{l^2 + m^2 + n^2 = 1}$ (1)

Also, $l = \cos \alpha$, $m = \cos \beta$ and $n = \cos \gamma$

Substituting in (1), we get

$$\boxed{\cos^2 \alpha + \cos^2 \beta + \cos^2 \gamma = 1}$$

We have, $\cos^2 \alpha = 1 - \sin^2 \alpha$, $\cos^2 \beta = 1 - \sin^2 \beta$, $\cos^2 \gamma = 1 - \sin^2 \gamma$

$\therefore$ $(1 - \sin^2 \alpha) + (1 - \sin^2 \beta) + (1 - \sin^2 \gamma) = 1$

$\therefore$ $\boxed{\sin^2 \alpha + \sin^2 \beta + \sin^2 \gamma = 2}$

DIRECTION RATIOS

If the direction cosines of a line are proportional to the numbers a, b, c then these are called proportional direction cosines or direction ratios of the line.

If the actual direction cosines are l, m, n then

$$\frac{l}{a} = \frac{m}{b} = \frac{n}{c} = \frac{\sqrt{l^2 + m^2 n^2}}{\sqrt{a^2 + b^2 + c^2}} = \frac{1}{\sqrt{a^2 + b^2 + c^2}}$$

$\therefore$ $l = \dfrac{a}{\sqrt{a^2 + b^2 + c^2}}$, $m = \dfrac{b}{\sqrt{a^2 + b^2 + c^2}}$, $n = \dfrac{c}{\sqrt{a^2 + b^2 + c^2}}$

or $l = \dfrac{a}{\sqrt{\Sigma a^2}}$, $m = \dfrac{b}{\sqrt{\Sigma a^2}}$, $n = \dfrac{c}{\Sigma a^2}$

where $\Sigma a^2 = a^2 + b^2 + c^2$.

Cor:

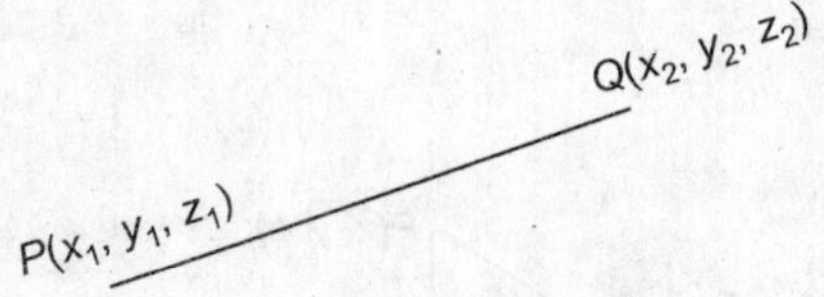

The direction cosines of the join of two points $P(x_1, y_1, z_1)$ and $Q(x_2, y_2, z_2)$ are proportional to $x_2 - x_1, y_2 - y_1, z_2 - z_1$

i.e. $x_2 - x_1, y_2 - y_1, z_2 - z_1$ are the direction ratios of the line.

ANGLE BETWEEN TWO LINES

If θ is the angle between two lines whose direction cosines are (l_1, m_1, n_1) and (l_2, m_2, n_2) prove that $\cos \theta = l_1 l_2 + m_1 m_2 + n_1 n_2$.

Proof: Let θ be the angle between the given lines. Through O draw $OP \parallel AB$ and $OQ \parallel CD$ and each of unit length so that $\angle POQ = \theta$.

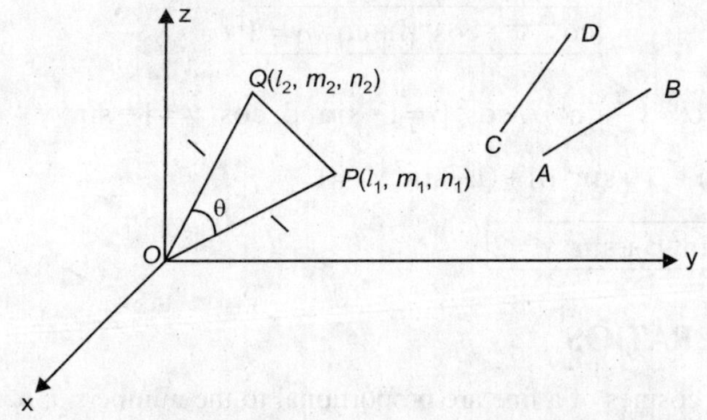

$\therefore$ The co-ordinates of P and Q are (l_1, m_1, n_1) and (l_2, m_2, n_2)

$$\therefore \quad PQ^2 = (l_1 - l_2)^2 + (m_1 - m_2)^2 + (n_1 - n_2)^2$$

$$= (l_1^2 + m_1^2 + n_1^2) + (l_2^2 + m_2^2 + n_2^2) - 2(l_1 l_2 + m_1 m_2 + n_1 n_2)$$

$$= 1 + 1 - 2(l_1 l_2 + m_1 m_2 + n_1 n_2)$$

From $\triangle POQ$ by cosine formula, we have

$$\cos \theta = \frac{OP^2 + OQ^2 - PQ^2}{2 \cdot OP \cdot OQ}$$

$$= \frac{1+1-[1+1-2(l_1 l_2 + m_1 m_2 + n_1 n_2)]}{2 \cdot 1 \cdot 1}$$

$$= l_1 l_2 + m_1 m_2 + n_1 n_2$$

$$\therefore \quad \cos\theta = l_1 l_2 + m_1 m_2 + n_1 n_2$$

$$\therefore \quad \theta = \cos^{-1}(l_1 l_2 + m_1 m_2 + n_1 n_2)$$

Cor 1: Prove that $\sin\theta = \pm\sqrt{\Sigma(m_1 n_2 - m_2 n_1)^2}$

Proof: We have

$$\sin^2\theta = 1 - \cos^2\theta$$

$$= (l_1^2 + m_1^2 + n_1^2)(l_2^2 + m_2^2 + n_2^2) - (l_1 l_2 + m_1 m_2 + n_1 n_2)^2$$

$$= (m_1 m_2 - m_2 n_1)^2 + (n_1 l_2 + n_2 l_1)^2 + (l_1 m_2 - m_1 l_2)^2$$

$$= \Sigma(m_1 n_2 - m_2 n_1)^2$$

$$\therefore \qquad \sin\theta = \pm\sqrt{\Sigma((m_1 n_2 - m_2 n_1)^2}$$

Cor 2: The two lines with direction cosines as l_1, m_1, n_1 and l_2, m_2, n_2 are perpendicular if

$$l_1 l_2 + m_1 m_2 + n_1 n_2 = 0 \qquad\qquad \because \cos 90° = 0$$

Cor 3: The two lines with direction cosines as l_1, m_1, n_1 and l_2, m_2, n_2 are parallel if

$$l_1 = l_2, \; m_1 = m_2, \; n_1 = n_2$$

Cor 4: The angle θ between two lines whose direction ratios are a_1, b_1, c_1 and a_2, b_2, c_2 is given by

(i)
$$\cos\theta = \frac{a_1 a_2 + b_1 b_2 + c_1 c_2}{\sqrt{a_1^2 + b_1^2 + c_1^2}\,\sqrt{a_2^2 + b_2^2 + c_2^2}}$$

(ii)
$$\sin\theta = \frac{\sqrt{(a_1 b_2 - a_2 b_1)^2 + (b_1 c_2 - b_2 c_1)^2 + (c_1 a_2 - c_2 a_1)^2}}{\sqrt{a_1^2 + b_1^2 + c_1^2}\,\sqrt{a_2^2 + b_2^2 + c_2^2}}$$

$$= \frac{\sqrt{\Sigma(a_1 b_2 - b_1 a_2)^2}}{\sqrt{\Sigma a_1^2}\,\sqrt{\Sigma a_2^2}}$$

where, actual direction cosines of the given lines are

$$\frac{a_1}{\sqrt{\Sigma a_1^2}}, \frac{b_1}{\sqrt{\Sigma a_1^2}}, \frac{c_1}{\sqrt{\Sigma a_1^2}} \text{ and } \frac{a_2}{\sqrt{\Sigma a_2^2}}, \frac{b_2}{\Sigma a_2^2}, \frac{c_2}{\sqrt{\Sigma a_2^2}}.$$

Cor 5: The condition that the lines whose direction ratios are a_1, b_1, c_1 and a_2, b_2, c_2 should be perpendicular be $a_1 a_2 + b_1 b_2 + c_1 c_2 = 0$ $\qquad (\because \cos 90° = 0)$

Cor 6: The condition that the lines whose direction ratios are a_1, b_1, c_1 and a_2, b_2, c_2 should be parallel

$$\frac{a_1}{a_2} = \frac{b_1}{b_2} = \frac{c_1}{c_2}.$$

Examples

(1) A line makes an angle of $45°$ with x-axis and $60°$ with y-axis. What angle does it make with z-axis?

Solution: The direction cosines of a line are $\cos \alpha$, $\cos \beta$ $\cos \gamma$.

A line makes $45°$ with x-axis $\therefore \cos 45° = \dfrac{1}{\sqrt{2}}$

A line makes $60°$ with y-axis $\therefore \cos 60° = \dfrac{1}{2}$

A line makes $\gamma°$ with z-axis $\cos \gamma = ?$

But we know that $\cos^2 \alpha + \cos^2 \beta + \cos^2 \gamma = 1$

$$\therefore \quad \frac{1}{2} + \frac{1}{4} + \cos^2 \gamma = 1$$

$$\therefore \quad \cos \gamma = \pm \frac{1}{2}$$

$$\gamma = 60 \text{ or } 120°$$

(2) Find the co-ordinates of the foot of the $\perp$ from $A(1, 1, 1)$ on the line joining $B(1, 4, 6)$ and $C(5, 4, 4)$.

Solution:

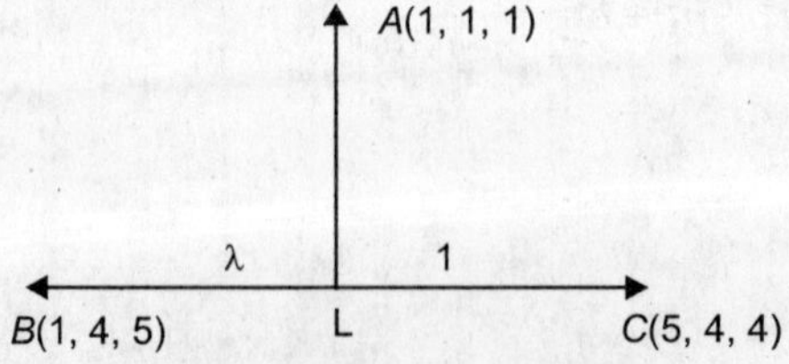

Let L be the foot of the $\perp$ from A on BC, divide BC in the ratio $\lambda : 1$ so that

$$L \equiv \left(\frac{5\lambda + 1}{\lambda + 1}, \frac{4\lambda + 4}{\lambda + 1}, \frac{4\lambda + 6}{\lambda + 1} \right) \qquad (1)$$

The direction ratios of *AL* are

$$\frac{5\lambda+1}{\lambda+1}-1,\ \frac{4\lambda+4}{\lambda+1}-1,\ \frac{4\lambda+6}{\lambda+1}-1$$

and the direction ratios of *BC* are

$$5-1,\ 4-4,\ 4-6 \qquad (\because\ x_2-x_1,\ y_2-y_1,\ z_2-z_1)$$

i.e.,   4, 0, –2.

since *AL* $\perp$ to *BC*, we have

$$aa' + bb' + cc' = 0$$

$$\left(\frac{5\lambda+1}{\lambda+1}-1\right)(4)+3\times0+\left(\frac{4\lambda+6}{\lambda+1}-1\right)(-2)=0$$

which gives $\lambda = 1$.

Substituting in (1), we get

$$L \equiv \left(\frac{5+1}{1+1},\ \frac{4+4}{1+1},\ \frac{4+6}{1+1}\right)=(3,\ 4,\ 5).$$

(3) Find the angle between the lines *OP* and *OQ* where *O* is the origin and the co-ordinates of *P* and *Q* are (1, 2, 3) and (1, 1, 4).

Solution: Let θ be the angle between *OP* and *OQ*. The direction ratios of *OP* are
$1 - 0, 2 - 0, 3 - 0$, i.e., 1, 2, 3

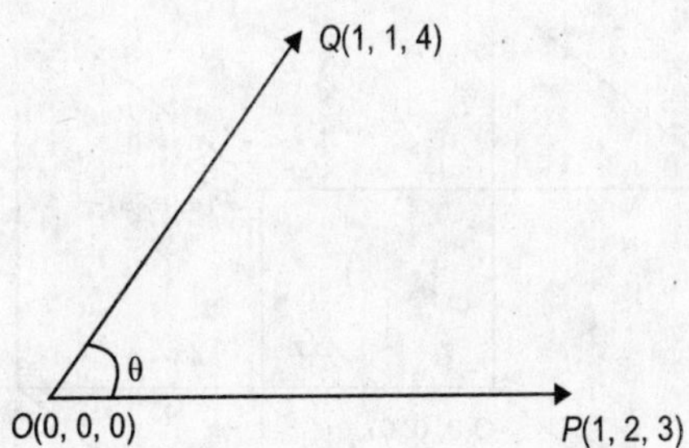

and the direction ratios of *OQ* are
$1 - 0, 1 - 0, 4 - 0$, i.e., 1, 1, 4

Then,
$$\cos\theta = \frac{a_1a_2+b_1b_2+c_1c_2}{\sqrt{a_1^2+b_1^2+c_1^2}\ \sqrt{a_2^2+b_2^2+c_2^2}}$$

$$= \frac{1\times1+2\times1+3\times4}{\sqrt{1^2+2^2+3^2}\ \sqrt{1^2+1^1+4^2}}$$

$$= \frac{1+2+12}{\sqrt{1+4+9}\sqrt{1+1+16}}$$

$$= \frac{15}{\sqrt{14}\sqrt{18}} = \frac{5}{\sqrt{2}\sqrt{14}}$$

$$\therefore \quad \theta = \cos^{-1}\left(\frac{5}{\sqrt{2}\sqrt{14}}\right) = \cos^{-1}\left(\frac{5}{2\sqrt{7}}\right)$$

Definition

(a) A parallelepiped is a solid bounded by three pairs of parallel plane faces.

(b) A rectangular parallelepiped is a parallelepiped having rectangular faces, i.e., a room.

(c) A cube is a parallelepiped having equal square faces.

(4) A line makes angles α, β, γ, δ with the diagonal of a cube. Prove that

$$\cos^2\alpha + \cos^2\beta + \cos^2\gamma + \cos^2\delta = \frac{4}{3}.$$

Proof: Let O, the corner of the cube, be the origin, and OA, OB, OC the edges through it as the axes. Let $OA = OB = OC = a$ then the co-ordinates of the corners are as shown in the figure. The four diagonals are OP, AA', BB' and CC'.

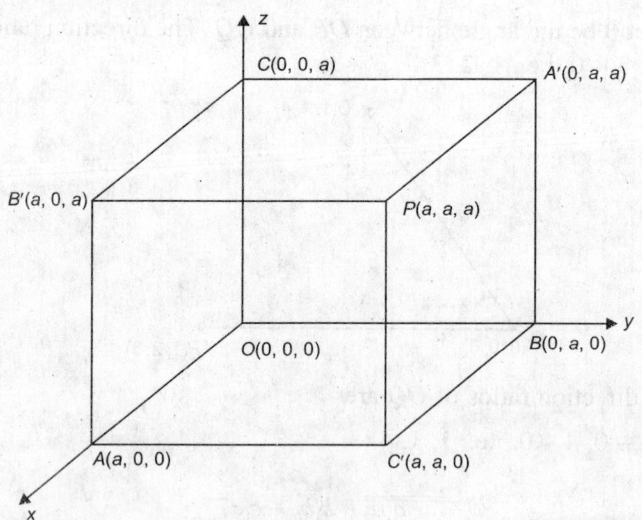

Clearly, the direction cosines of OP are

$$\frac{a-0}{\sqrt{a^2+a^2+a^2}},\ \frac{a-0}{\sqrt{a^2+a^2+a^2}},\ \frac{a-0}{\sqrt{a^2+a^2+a^2}}\ \text{i.e.,}\ \frac{1}{\sqrt{3}},\ \frac{1}{\sqrt{3}},\ \frac{1}{\sqrt{3}}.$$

Similarly, the direction cosines of AA' are $\dfrac{-1}{\sqrt{3}}, \dfrac{1}{\sqrt{3}}, \dfrac{1}{\sqrt{3}}$

The direction cosines of BB' are $\dfrac{1}{\sqrt{3}}, \dfrac{-1}{\sqrt{3}}, \dfrac{1}{\sqrt{3}}$ and the direction cosines of

CC' are $\dfrac{1}{\sqrt{3}}, \dfrac{1}{\sqrt{3}}, \dfrac{-1}{\sqrt{3}}$. Let l, m, n be the direction cosines of the given line which

makes angle α, β, γ and δ with OP, AA', BB' and CC' respectively.
   Then,

$$\cos\alpha = \frac{1}{\sqrt{3}}(l+m+n), \quad \cos\beta = \frac{1}{\sqrt{3}}(-l+m+n)$$

$$\cos\gamma = \frac{1}{\sqrt{3}}(l-m+n), \quad \cos\delta = \frac{1}{\sqrt{3}}(l+m-n)$$

Squaring and adding, we get

$$\cos^2\alpha + \cos^2\beta + \cos^2\gamma + \cos^2\delta = \frac{1}{3}\Big[(l+m+n)^2 + (-l+m+n)^2$$

$$+(l-m+n)^2 + (l+m-n)^2\Big]$$

$$= \frac{4}{3}(l^2+m^2+n^2) = \frac{4}{3}$$

$$\therefore \quad \cos^2\alpha + \cos^2\beta + \cos^2\gamma + \cos^2\delta = \frac{4}{3}. \qquad\qquad (\because l^2+m^2+n^2=1)$$

Cor: We have

$$\cos^2\alpha + \cos^2\beta + \cos^2\gamma + \cos^2\delta = \frac{4}{3}.$$

$$\therefore \quad 1-\sin^2\alpha + 1-\sin^2\beta + 1-\sin^2\gamma + 1-\sin^2\delta = \frac{4}{3}.$$

$$\therefore \quad \sin^2\alpha + \sin^2\beta + \sin^2\gamma + \sin^2\delta = 4 - \frac{4}{3} = \frac{8}{3}.$$

$$\therefore \quad \sin^2\alpha + \sin^2\beta + \sin^2\gamma + \sin^2\delta = \frac{8}{3}.$$

(5) Show that the angle between two diagonals of a cube is $\cos^{-1}\left(\dfrac{1}{3}\right)$.

Solution: Take O, the corner of the cube, as origin and OA, OB, OC the edges through it as the axes. Let $OA = OB = OC = a$. Then the co-ordinates of the corners are as shown in the figure. The four diagonals are OP, AA', BB' and CC'.

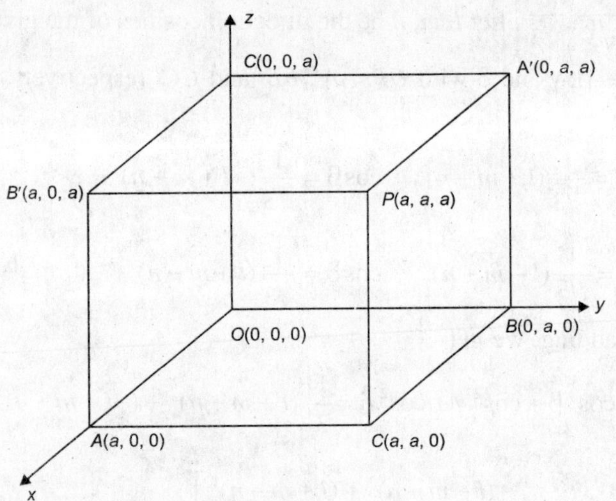

Clearly, the direction cosines of OP are

$$\frac{a-0}{\sqrt{a^2+a^2+a^2}}, \; \frac{a-0}{\sqrt{a^2+a^2+a^2}}, \; \frac{a-0}{\sqrt{a^2+a^2+a^2}} \text{ i.e., } \frac{1}{\sqrt{3}}, \frac{1}{\sqrt{3}}, \frac{1}{\sqrt{3}}.$$

Similarly, the direction cosines of BB' are $\dfrac{1}{\sqrt{3}}, \dfrac{-1}{\sqrt{3}}, \dfrac{1}{\sqrt{3}}$.

Let θ be the angle between OP and BB'
Then

$$\cos\theta = l_1 l_2 + m_1 m_2 + n_1 n_2$$

$$= \frac{1}{\sqrt{3}}\times\frac{1}{\sqrt{3}} + \frac{1}{\sqrt{3}}\times\frac{-1}{\sqrt{3}} + \frac{1}{\sqrt{3}}\times\frac{1}{\sqrt{3}}$$

$$= \frac{1}{3} - \frac{1}{3} + \frac{1}{3} = \frac{1}{3}$$

$$\therefore \quad \theta = \cos^{-1}\left(\frac{1}{3}\right)$$

(6) Show that the angles between four diagonals of a rectangular parallelepiped are:

$$\cos^{-1}\left(\frac{\pm a^2 \pm b^2 \pm c^2}{a^2 + b^2 + c^2}\right)$$

Solution: A rectangular parallelepiped is shown in the figure with O as the origin and three rectangular faces through it as the three rectangular co-ordinates planes.

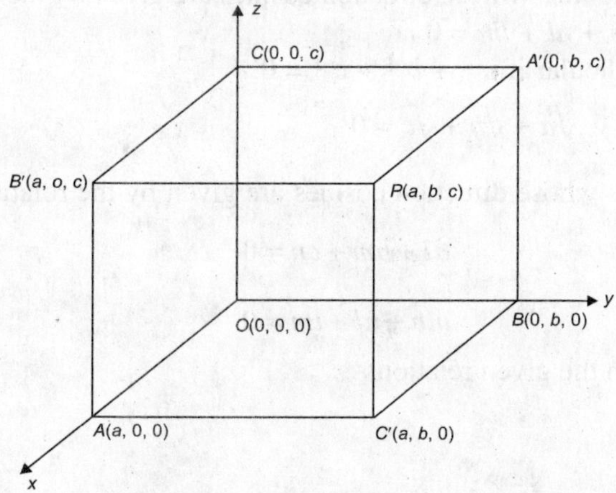

Let $OA = a$, $OB = b$ and $OC = c$. Then the co-ordinates of the corners are as given in the figure. The four diagonals are OP, AA', BB' and CC'.

Now, the direction cosines of OP are $\dfrac{a}{\sqrt{\Sigma a^2}}, \dfrac{b}{\sqrt{\Sigma a^2}}, \dfrac{c}{\sqrt{\Sigma a^2}}$

and the direction cosines of BB' are $\dfrac{a}{\sqrt{\Sigma a^2}}, \dfrac{-b}{\sqrt{\Sigma a^2}}, \dfrac{c}{\sqrt{\Sigma a^2}}$

Let θ be the angle between OP and BB'

Then $\cos\theta = \dfrac{a}{\sqrt{\Sigma a^2}} \cdot \dfrac{a}{\sqrt{\Sigma a^2}} + \dfrac{b}{\sqrt{\Sigma a^2}} \cdot \dfrac{(-b)}{\sqrt{\Sigma a^2}} + \dfrac{c}{\sqrt{\Sigma a^2}} \cdot \dfrac{c}{\sqrt{\Sigma a^2}}$

$$= \frac{a^2 - b^2 + c^2}{a^2 + b^2 + c^2}$$

where $\Sigma a^2 = a^2 + b^2 + c^2$

Similarly, we can find the angles between the other pairs of diagonals. In general, we can write

$$\cos\theta = \frac{\pm a^2 \pm b^2 \pm c^2}{a^2 + b^2 + c^2}$$

$$\therefore \quad \theta = \cos^{-1}\left(\frac{\pm a^2 \pm b^2 \pm c^2}{a^2 + b^2 + c^2}\right)$$

(7) Prove that the lines whose direction cosines are given by the relations $al + bm + cn = 0$ and $mn + nl + lm = 0$ are

(i) Perpendicular if $a^{-1} + b^{-1} + c^{-1} = 0$

(ii) Parallel if $\sqrt{a} + \sqrt{b} + \sqrt{c} = 0$

Proof: The lines whose direction cosines are given by the relations

$$al + bm + cn = 0$$

$$mn + nl + lm = 0$$

Eliminate n from the given relations

We have

$$(m + l)\left(\frac{-al - bm}{c}\right) + lm = 0 \qquad \left(\because n = \frac{-al - bm}{c}\right)$$

$$\therefore \quad al^2 + (c - a - b)lm + bm^2 = 0$$

$$\therefore \quad a\left(\frac{l}{m}\right)^2 + (c - a - b)\left(\frac{l}{m}\right) + b = 0 \qquad (1)$$

which is a quadratic equation in $\dfrac{l}{m}$.

If l_1, m_1, n_1 and l_2, m_2, n_2 are the direction cosines of these lines then $\dfrac{l_1}{m_1}, \dfrac{l_2}{m_2}$ are the roots of (1).

$$\therefore \quad \frac{l_1}{m_1} \cdot \frac{l_2}{m_2} = \frac{b}{a}.$$

or $$\frac{l_1 l_2}{\dfrac{1}{a}} = \frac{m_1 m_2}{\dfrac{1}{b}} = \frac{n_1 n_2}{\dfrac{1}{c}} = k \text{ (say)} \qquad\qquad \text{by symmetry}$$

The lines with be $\perp$ if

$$l_1 l_2 + m_1 m_2 + n_1 n_2 = k\left(\frac{1}{a} + \frac{1}{b} + \frac{1}{c}\right) = 0$$

i.e., if $\dfrac{1}{a} + \dfrac{1}{b} + \dfrac{1}{c} = 0$ i.e., $a^{-1} + b^{-1} + c^{-1} = 0$.

The lines will be parallel if

$$l_1 = l_2,\ m_1 = m_2,\ n_1 = n_2$$

i.e., if
$$\frac{l_1}{m_1} = \frac{l_2}{m_2}$$

i.e., if
$$(c - b - a)^2 = 4ab.$$

$\therefore$
$$c - b - a = \pm 2\sqrt{ab}$$

$$c = a + b \pm \sqrt{ab} = \left(\sqrt{a} + \sqrt{b}\right)^2$$

$\therefore$
$$\pm\sqrt{c} = \sqrt{a} \pm \sqrt{b}$$

$\therefore$
$$\sqrt{a} + \sqrt{b} + \sqrt{c} = 0.$$

Projection: The projection of a line segment of a line is the line joining the feet of the perpendicular from its ends on the line.

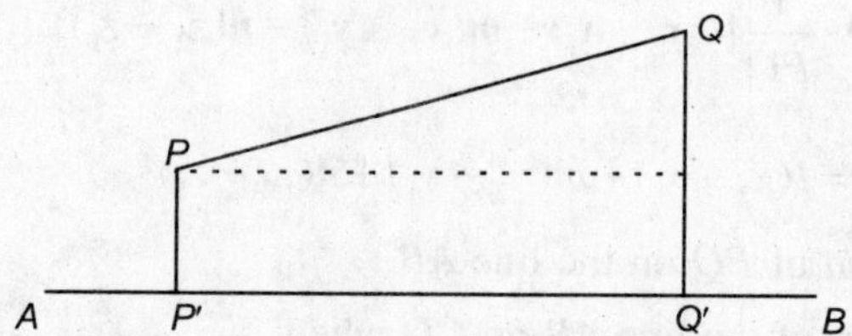

Thus, $P'Q'$ is the projection of PQ on the line AB.

Theorem: Prove that the projection of the join of two points $P(x_1, y_1, z_1)$, $Q(x_2, y_2, z_2)$ on a line AB whose direction cosines are l, m, n is

$$l(x_2 - x_1) + m(y_2 - y_1) + n(z_2 - z_1)$$

Proof: Let $P(x_1, y_1, z_1)$, $Q(x_2, y_2, z_2)$ be given points and AB the line whose direction cosines are l, m, n. Draw PP', $QQ' \perp$ on AB.

Then $P'Q' = PQ \cos\theta$ (1) where θ is the angle which PQ makes with AB.

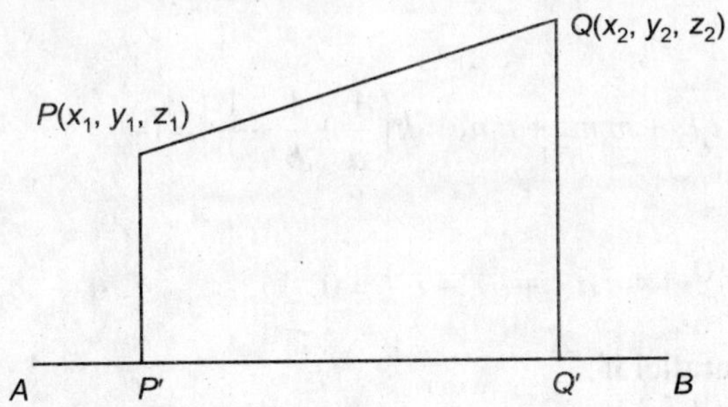

Now d.r.s. of PQ are $x_2 - x_1,\ y_2 - y_1,\ z_2 - z_1$

$\therefore$ d.c.s. of PQ are

$$\frac{x_2 - x_1}{PQ},\ \frac{y_2 - y_1}{PQ},\ \frac{z_2 - z_1}{PQ}$$

where

$$PQ = \sqrt{(x_2 - x_1)^2 + (y_2 - y_1)^2 + (z_2 - z_1)^2}$$

Also d.c.s. of AB are $l,\ m,\ n$

$\therefore$

$$\cos\theta = l\left(\frac{x_2 - x_1}{PQ}\right) + m\frac{(y_2 - y_1)}{PQ} + n\frac{(z_2 - z_1)}{PQ}$$

$\therefore$

$$\cos\theta = \frac{1}{PQ}[l(x_2 - x_1) + m(y_2 - y_1) + n(z_2 - z_1)]$$

Substituting this value in (1), we get

$$P'Q' = PQ \cdot \frac{1}{PQ}[l(x_2 - x_1) + m(y_2 - y_1) + n(z_2 - z_1)]$$

$\therefore$

$$P'Q' = l(x_2 - x_1) + m(y_2 - y_1) + n(z_2 - z_1)$$

which is the projection of PQ on the line AB.

(8) Find the projection of the line AB on CD where

$$A = (1, 2, 3),\ B = (-1, 0, 2),\ C = (1, 4, 2),\ D = (2, 0, -1)$$

Solution:

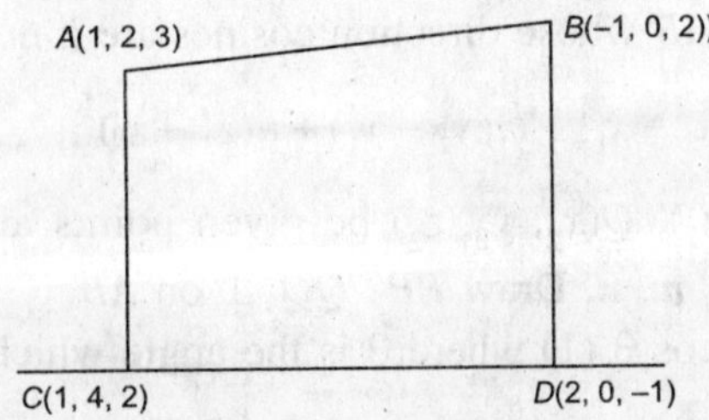

First calculate the direction cosines of the line CD

i.e. $$\frac{2-1}{\sqrt{(2-1)^2+(0-4)^2+(-1,-2)^2}},\ \frac{0-4}{\sqrt{(2-1)^2+(0-4)^2+(-1-2)]^2}}$$

$$\frac{-1-2}{\sqrt{(2-1)^2+(0-4)^2+(-1-2)^2}}$$

$\therefore\ \dfrac{1}{\sqrt{26}},\ \dfrac{-4}{\sqrt{26}},\ \dfrac{-3}{26}$ are the d.c.s. of the line CD.

The projection of the line AB on CD is

$$A'B' = l(x_2 - x_1) + m(y_2 = y_1) + n(z_2 - z_1)$$

$$= \frac{1}{\sqrt{26}}(-1-1) + \left(\frac{-4}{\sqrt{26}}\right)(0-2) + \left(\frac{-3}{\sqrt{26}}\right)(2-3)$$

$$= \frac{1}{\sqrt{26}}(-2+8+3)$$

$$= \frac{9}{\sqrt{26}}$$

QUESTION BANK 1

(1) Define the direction cosines of a striaght line. If l, m, n are the direction cosines of a line prove that $l^2 + m^2 + n^2 = 0$.

(2) A line makes an angle of $45°$ with OX and $60°$ with OY, what angle does it make with OZ?

(3) Find the angle between two lines whose direction cosines are (l_1, m_1, n_1) and (l_2, m_2, n_2).

(4) A line makes angles α, β, γ, δ with diagonals of a cube. Prove that

$$\cos^2 \alpha + \cos^2 \beta + \cos^2 \gamma + \cos^2 \delta = \frac{4}{3}.$$

(5) Find the projection of the line PQ on AB where
$P = (1, 2, 3)$, $Q = -1, 0, 2)$, $A = 1, 4, 2)$ $B = (2, 0, -1)$

(6) Find the angle between two lines whose direction cosines are given by $l + 3m + 5n = 0$ and $2mn - 6nl - 5lm = 0$.

(7) Find the angle between any two diagonals of a cube.

(8) Find the co-ordinates of the foot of the perpendicular from $A(1, 1, 1)$ to the line joining $B(1, 4, 6)$ and $C(5, 4, 4)$.

(9) If $A(4, 3, 2)$, $B(5, 4, 6)$, $C(-1, -1, 5)$ are the corners of a triangle, find the co-ordinates of the point in which the bisector of the angle A meets the side BC.

(10) Find the ratio in which XOY plane divides the line joining of the points $(-3, 4, 8)$ and $(5, -6, 4)$ and thus write the co-ordinates of the point of division.

(11) Find the angle between the lines whose direction cosines are proportional to 1, 2, 1 and 2, -3, 6.

(12) Find the direction cosines of a line perpendicular to the two lines whose direction ratios are 1, 2, 3 and -2, 1, 4.

(13) $A(1, 8, 4)$, $B(0, -1, 1)$, $C(2, -3, 1)$ are three points and D is the foot of the perpendicular from A on BC. Find the co-ordinates of D.

Ans. $D(4, 5, -2)$

(14) Show that the points $(1, 3, 4)$, $(-1, 6, 10)$, $(-7, 4, 7)$ and $(-5, 1, 1)$ are the vertices of a rhombus.

(15) Show that the points $(1, 2, 3)$, $(-1, -2, -1)$, $(2, 3, 2)$ and $(4, 7, 6)$ are the vertices of a parallelogram.

UNIT 2

Plane

DEFINITION

A plane is a surface such that the line joining any two of its points lies wholly on the surface.

(1) Equation of a Plane

Let ABC be the plane. Draw $OK \perp$ to the plane. Let $OK = P$ and its direction cosines be l, m, n.

Let $P\,(x, y, z)$ be any point on the plane. Join OP and PK. Then

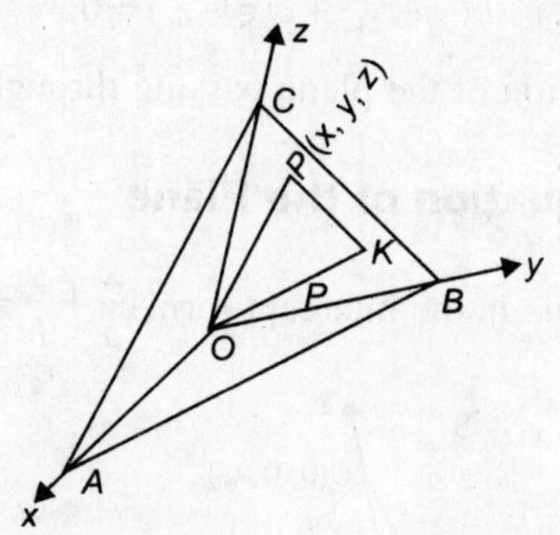

$\qquad OK$ = projection of OP on OK.

$$p = l(x - o) + m(y - o) + n(z - o)$$

$$\therefore \qquad\qquad\qquad lx + my + nz = p \qquad\qquad\qquad (1)$$

which is the equation of a plane in the *normal form*. This shows that the general equation of a plane is of the form

$$ax + by + cz + d = 0 \qquad\qquad\qquad (2)$$

i.e., every equation of first degree in x, y, z represents a plane.

Comparing the coefficients of (1) and (2), we get,

$$\frac{l}{a} = \frac{m}{b} = \frac{n}{c}$$

i.e., the direction cosines of the normal to the plane (2) are proportional to a, b, c.

Thus, the coefficients of x, y, z in the equation of any plane are the direction ratios of the normal to the plane.

(2) Equation of a Plane through the Point (x_1, y_1, z_1)

Any plane through the point (x_1, y_1, z_1) is

$$a(x - x_1) + b(y - y_1) + c(z - z_1) = 0$$

For, let the equation of the plane be

$$ax + by + cz + d = 0$$

Since it passes through the point (x_1, y_1, z_1)

$$ax_1 + by_1 + cz_1 + d = 0$$

Subtracting, we get

$$a(x - x_1) + b(y - y_1) + c(z - z_1) = 0$$

which is the required equation of the plane passing through the point (x_1, y_1, z_1)

(3) Intercept form Equation of the Plane

To find the equation of plane in the intercept form $\dfrac{x}{a} + \dfrac{y}{b} + \dfrac{z}{c} = 1$

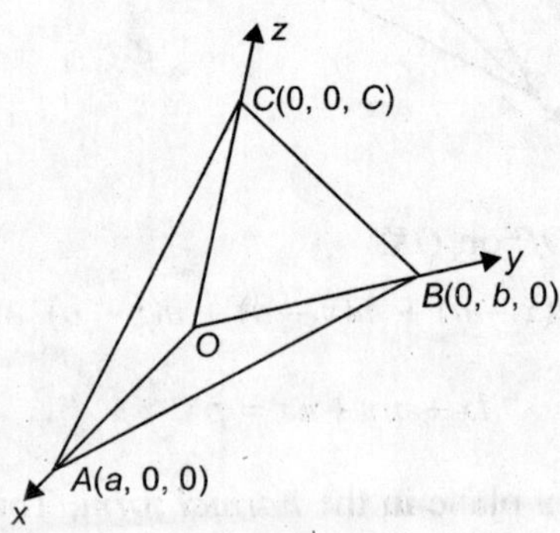

Let the required equation of the plane be

$$\alpha x + \beta y + \gamma z + \delta = 0 \tag{1}$$

The plane cuts the axes at A, B, C such that $OA = a$, $OB = b$, $OC = c$, i.e. it passes through the points $A(a, 0, 0)$ $B(0, b, 0)$ and $C(0, 0, c)$

$$\therefore \quad \alpha a + \delta = 0;\ \beta b + \delta = 0;\ \gamma c + \delta = 0$$

$$\therefore \quad \alpha = \frac{-\delta}{a},\ \beta = \frac{-\delta}{b},\ \gamma = \frac{-\delta}{c}$$

Substituting these values in (1), we get

$$\frac{-\delta}{a}x - \frac{\delta}{b}y - \frac{\delta}{c}z + \delta = 0$$

$$\therefore \quad \boxed{\frac{x}{a} + \frac{y}{b} + \frac{z}{c} = 1}$$

which is the equation of a plane in the intercept form.

(4) Plane Passing through Three Points

Let the equation plane be

$$ax + by + cz + d = 0 \tag{1}$$

This plane passes through the points $P(x_1, y_1, z_1)$, $Q(x_2, y_2, z_2)$, and $R(x_3, y_3, z_3)$.
Since (1) passes through the points P, Q, R, we have

$$ax_1 + by_1 + cz_1 + d = 0$$

$$ax_2 + by_2 + cz_2 + d = 0$$

$$ax_3 + by_3 + cz_3 + d = 0$$

Eliminating a, b, c from these equations, we get

$$\begin{vmatrix} x - x_1 & y - y_1 & z - z_1 \\ x_2 - x_1 & y_2 - y_1 & z_2 - z_1 \\ x_3 - x_1 & y_3 - y_1 & z_3 - z_1 \end{vmatrix} = 0$$

This is the equation of the plane passing through three points.

(5) Condition for Four Points to be Coplanar

Let $P(x_1, y_1, z_1)$, $Q(x_2, y_2, z_2)$, and $R(x_3, y_3, z_3)$ and $S(x_4, y_4, z_4)$ be four points in the plane. Then the condition for these four points to be coplanar is

$$\begin{vmatrix} x_2 - x_1 & y_2 - y_1 & z_2 - z_1 \\ x_3 - x_1 & y_3 - y_1 & z_3 - z_1 \\ x_4 - x_1 & y_4 - y_1 & z_4 - z_1 \end{vmatrix} = 0$$

ANGLE BETWEEN TWO PLANES

Definition

The angle between two planes is equal to the angle between their normals.

Let the two planes be

$$ax + by + cz + d = 0$$

and

$$a_1 x + b_1 y + c_1 z + d_1 = 0$$

Now, the direction ratios of their normals are a, b, c and a_1, b_1, c_1.

If θ be the angle between them, then

$$\cos\theta = \frac{aa_1 + bb_1 + cc_1}{\sqrt{a^2 + b^2 + c^2}\,\sqrt{a_1^2 + b_1^2 + c_1^2}}$$

Cor 1: The two planes will be $\perp$ if

$$aa_1 + bb_1 + cc_1 = 0 \qquad\qquad \because \cos 90° = 0$$

Cor 2: The two planes will be parallel if

$$\frac{a}{a_1} = \frac{b}{b_1} = \frac{c}{c_1}$$

Cor 3: Any plane parallel to the plane

$$ax + by + cz + d = 0 \ is \ ax + by + cz + k = 0 \qquad (k \text{ is constant})$$

The direction ratios of their normals are the same.

Examples

(1) Find the angle between the planes $3x - 2y + 6z - 1 = 0$ and $2x + 2y - z + 3 = 0$

Solution: The angle between two planes is the angle between their normals.

Given: $\qquad\qquad 3x - 2y + 6z - 1 = 0$ and

$2x + 2y - z + 3 = 0$ are the equations of the planes.

The direction ratios of their normals are

$$3, -2, 6 \text{ and } 2, 2, -1$$

If θ be the angle between them, then

$$\cos\theta = \frac{aa_1 + bb_1 + cc_1}{\sqrt{a^2 + b^2 + c^2}\,\sqrt{a_1^2 + b_1^2 + c_1^2}}$$

$$= \frac{(3)(2) + (-2)(2) + (6)(-1)}{\sqrt{9+4+36}\sqrt{4+4+1}}$$

$$= \frac{6-4-6}{\sqrt{49}\sqrt{9}} = \frac{4}{7\times 3} = \frac{-4}{21}$$

$$\therefore \quad \theta = \cos^{-1}\left(\frac{-4}{21}\right)$$

(2) Find the equation of the plane which bisects the straight line joining the points $A(1, 3, -2)$ and $B(3, 1, 6)$ at right angles.

Solution: Given $A(1, 3, -2)$ and $B(3, 1, 6)$ are two points.

$\therefore$ Midpoint of AB is

$$C = \left(\frac{1+3}{2}, \frac{3+1}{2}, \frac{-2+6}{2}\right) = (2, 2, 2).$$

Now, direction ratios of the line AB are

$$3 - 2, 1 - 3, 6 + 2 \text{ i.e., } 2, -2, 8 \text{ or } 1, -1, 4$$

Since the plane is $\perp$ to AB, normal to the plane is parallel to AB.

Hence, direction ratios of the normal to the plane are $(a, b, c) \equiv (1, -1, 4)$ and it passes through the point $(2, 2, 2)$.

$\therefore$ Required equation of the plane is

$$a(x - x_1) + b(y - y_1) + c(z - z_1) = 0$$

$$\therefore \quad 1(x - 2) - 1(y - z) + y(z - 2) = 0$$

i.e., $x - y + 4z - 8 = 0$

(3) Find the equation of the plane which passes through the points $(0, 1, 1)$, $(1, 1, 2)$ and $(-1, 2, -2)$.

Solution: The equation of the plane passing through three points is

$$\begin{vmatrix} x - x_1 & y - y_1 & z - z_1 \\ x_2 - x_1 & y_2 - y_1 & z_2 - z_1 \\ x_3 - x_1 & y_3 - y_1 & z_3 - z_1 \end{vmatrix} = 0$$

$$\therefore \quad \begin{vmatrix} x - 0 & y - 1 & z - 1 \\ 1 - 0 & 1 - 1 & 2 - 1 \\ -1 - 0 & 2 - 1 & -2 - 1 \end{vmatrix} = \begin{vmatrix} x & y - 1 & z - 1 \\ 1 & 0 & 1 \\ -1 & 1 & -3 \end{vmatrix} = 0$$

i.e., $x - 2y - z + 3 = 0$ is the required equation of the plane.

(4) Find the equation of the plane which passes through the point $(3, -3, 1)$ and is parallel to the plane $2x + 3y + 5z + 6 = 0$

Solution: Any plane parallel to the given plane is

$$2x + 3y + 5z + k = 0$$

This plane passes through $(3, -3, 1)$

$$\therefore \quad 2 \times 3 + 3 \times -3 + 5 \times 1 + k = 0 \quad \therefore \quad k = -2.$$

Hence, the required plane is

$$2x + 3y + 5z - 2 = 0$$

(5) Find the equation of the plane which passes through the point $(3, -3, 1)$ and is normal to the line joining the points $(3, 2, -1)$ and $(2, -1, 5)$.

Solution: Any plane through $(3, -3, 1)$ is

$$a(x - 3) + b(y + 3) + c(z - 1) = 0 \tag{1}$$

The direction ratios of the line joining the points $(3, 2, -1)$ and $(2, -1, 5)$ are

$$x_2 - x_1,\ y_2 - y_1,\ z_2 - z_1$$
$$2 - 3,\ -1 - 2,\ 5 + 1$$
$$-1,\ -3,\ 6 \text{ or } 1,\ 3,\ -6$$

This line is normal to the plane (1).

$\therefore$ a, b, c are proportional to $1, 3, -6$.

Substituting in (1)

$$1(x - 3) + 3(y + 3) - 6(z - 1) = 0$$

$$\therefore \qquad x + 3y - 6z + 12 = 0$$

(6) Find the equation of the plane which passes through $(3, -3, 1)$ and is perpendicular to the planes $7x + y + 2z = 6$ and $3x + 5y - 6z = 8$.

Solution: Any plane passing through $(-3, -3, 1)$ is

$$a(x - 3) + b(x + 3) + c(z - 1) = 0 \tag{1}$$

Plane (1) is perpendicular to the planes

$$7x + y + 2z = 6 \text{ and } 3x + 5y - 6z = 8$$

$$\therefore \qquad 7a + b + 2c = 0 \qquad\qquad (\because aa_1 + bb_1 + cc_1 = 0)$$

$$3a + 5b - 6c = 0$$

By cross multiplication

$$\frac{a}{\begin{vmatrix} 1 & 2 \\ 5 & -6 \end{vmatrix}} = \frac{b}{\begin{vmatrix} 2 & 7 \\ -6 & 3 \end{vmatrix}} = \frac{c}{\begin{vmatrix} 7 & 1 \\ 3 & 5 \end{vmatrix}}$$

i.e.,
$$\frac{a}{-16} = \frac{b}{-3} = \frac{c}{-2}$$

Hence, the required equation of the plane is

$$1(x-3) - 3(y+3) - 2(z-1) = 0$$

$$\therefore \qquad x - 3y - 2z - 10 = 0.$$

(7) Find the equation to the plane through $(-1, 3, 2)$ and perpendicular to the planes $x + 2y + 2z = 5$ and $3x + 3y + 2z = 8$.

Solution: Any plane through $(-1, 3, 2)$ is

$$a(x+1) + b(y-3) + c(z-2) = 0 \qquad\qquad (1)$$

Since (1) is $\perp$ to the planes

$$x + 2y + 2z = 5 \text{ and } 3x + 3y + 2z = 8$$

$$\therefore \qquad a + 2b + 2c = 0$$

$$\therefore \qquad 3a + 3b + 2c = 0$$

$$\therefore \qquad \frac{a}{4-6} = \frac{b}{6-2} = \frac{c}{3-6} \text{ or } \frac{a}{2} = \frac{b}{-4} = \frac{c}{3}$$

Putting these values in (1), we get

$$2(x+1) - 4(4-3)3(z-2) = 0$$

$$\therefore \qquad 2x - 4y - 3z + 8 = 0.$$

(8) Find the image (reflection) of the point $(2, -1, 3)$ in the plane

$$3x - 2y - z - 9 = 0.$$

Solution: Let $A = (2, -1, 3)$ be the given point and $B(x_1, y_1, z_1)$ be its image in the plane.

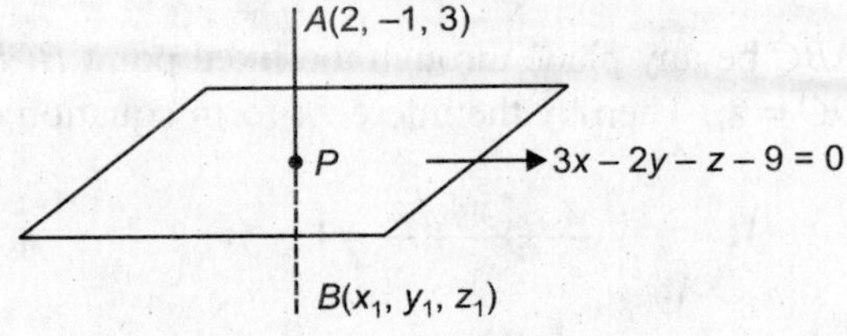

Then AB is $\perp$ to the plane and the midpoint P of AB must lie in the plane (1)

$$\therefore \quad \text{midpoint} \qquad P = \left(\frac{x_1 + 2}{2}, \frac{y_1 - 1}{2}, \frac{z_1 + 3}{2} \right)$$

Since P lies in the plane (1), co-ordinates of P satisfy the equation (1), i.e.

$$3\left(\frac{x_1 + 2)}{2} \right) - 2\left(\frac{y_1 - 1}{2} \right) - \left(\frac{z_1 + 3}{2} \right) - 9 = 0$$

$$\therefore \qquad\qquad 3x_1 - 2y_1 - z_1 - 13 = 0 \qquad\qquad (2)$$

The direction ratios of AB are

$$x_1 - 2, \ y_1 + 1, \ z_1 - 3$$

Since AB is $\perp$ to the plane, direction ratios of AB are proportional to the direction ratios of the normal to the plane, viz., $(3, -2, -1)$

i.e., $\dfrac{x_1 - 2}{3} = \dfrac{y_1 + 1}{-2} = \dfrac{z_1 - 3}{-1} = k$ (say)

$$\therefore \qquad\qquad x_1 = 2 + 3k, \ y_1 = -1 - 2k, \ z_1 = 3 - k \qquad\qquad (3)$$

From (2), $3(2 + 3k) - 2(-1 - 2k) - (3 - k) - 13 = 0$

$$\therefore \qquad\qquad 14k = 8 \ \therefore \ k = \frac{4}{7}.$$

Substituting in (3), we get

$$x_1 = \frac{26}{7}, \ y_1 = \frac{-15}{7}, \ z_1 = \frac{17}{7}.$$

$\therefore$ the reflection of A is $B\left(\dfrac{26}{7}, \dfrac{-15}{7}, \dfrac{17}{7} \right)$

(9) A variable plane passes through the fixed point (a, b, c) and meets the co-ordinates axes in A, B, C show that the locus of the point common to the co-ordinate plane is $\dfrac{a}{x} + \dfrac{b}{y} + \dfrac{c}{z} = 1$

Solution: Let ABC be any plane through the fixed point $H(a, b, c)$ such that $OA = x_1$, $OB = y_1$, $OC = z_1$. Then by the intercept form equation of the plane is

$$\frac{x}{x_1} + \frac{y}{y_1} + \frac{z}{z_1} = 1 \qquad\qquad (1)$$

Since this passes through (a, b, c), we have

$$\frac{a}{x_1} + \frac{b}{y_1} + \frac{c}{z_1} = 1 \tag{2}$$

The planes through A, B, C parallel to the co-ordinate planes are $x = x_1$, $y = y_1$, $z = z_1$ which meet in $P(x_1, y_1, z_1)$.

Therefore, changing x_1 to x, y_1 to y, z_1 to z in (2), the locus of P is

$$\frac{a}{x} + \frac{b}{y} + \frac{c}{z} = 1$$

PERPENDICULAR DISTANCE OF THE POINT (x_1, y_1, z_1) FROM THE PLANE $ax + by + cz + d = 0$

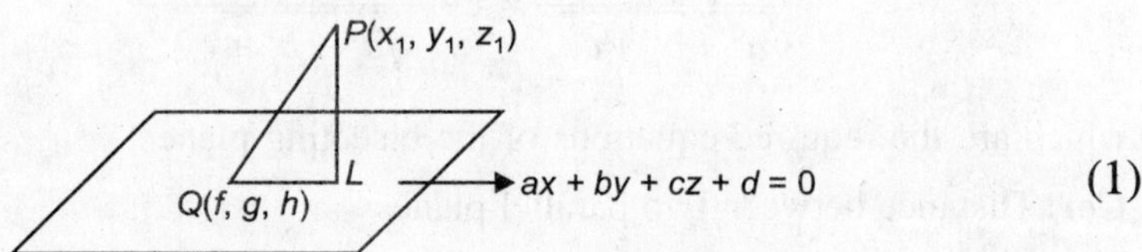

$$\tag{1}$$

Let PL be the required $\perp$ distance of $P(x_1, y_1\, z_1)$ from the plane

$$ax + by + cz + d = 0$$

So that its direction ratios are a, b, c.

$\therefore$ The actual direction cosines of PL are

$$\frac{a}{\sqrt{a^2 + b^2 + c^2}}, \frac{b}{\sqrt{a^2 + b^2 + c^2}}, \frac{c}{\sqrt{a^2 + b^2 + c^2}}$$

Let $Q(f, g, h)$ be a point on a plane (1)

$$\therefore \qquad af + bg + ch + d = 0 \tag{2}$$

$\therefore$ Projection of PQ on PL

$$= \frac{a(x_1 - f) + b(y_1 - g) + c(z_1 - h)}{\sqrt{a^2 + b^2 + c^2}}$$

$$\therefore \qquad PL = \frac{ax_1 + by_1 + cz_1 - (af + bg + ch)}{\sqrt{a^2 + b^2 + c^2}}$$

But $af + bg + ch = -d$

$$\therefore \qquad PL = \frac{ax_1 + by_1 + cz_1 + d}{\sqrt{a^2 + b^2 + c^2}}$$

which is the $\perp$ distance of any point in the space from the plane.

BISECTING PLANES

Let
$$ax + by + cz + d = 0 \qquad\qquad (1)$$

and
$$a_1 x + b_1 y + c_1 z + d_1 = 0 \qquad\qquad (2)$$

be the given planes. Let $P(x, y, z)$ be any point on either of the planes bisecting the angles between the planes (1) and (2). Then

$\perp$ distance of p from (1) = $\perp$ distance of P from (2)

$$\therefore \qquad \frac{ax + by + cz + d}{\sqrt{a^2 + b^2 + c^2}} = \pm\, \frac{a_1 x + b_1 y + c_1 z + d_1}{\sqrt{a_1^2 + b_1^2 + c_1^2}}$$

which are the required equations of the bisecting planes.

Cor: Distance between two parallel planes

$$ax + by + cz + d = 0 \text{ and } ax + by + cz + d_1 = 0 \text{ is}$$

$$= \frac{d - d_1}{\sqrt{a^2 + b^2 + c^2}}$$

Examples

(1) Find the perpendicular distance of the point $(3, 2, 1)$ from the plane $x + 2y + 3z + 4 = 0$.

Solution: The perpendicular distance of the point (x_1, y_1, z_1) from the plane $ax + by + cz + d = 0$ is

$$PL = \frac{ax_1 + by_1 + cz_1 + d}{\sqrt{a^2 + b^2 + c^2}}$$

$$= \frac{1 \times 3 + 2 \times 2 + 3 \times 3 + 4}{\sqrt{1^2 + 2^2 + 3^2}} = \frac{20}{\sqrt{14}}$$

(2) Find the equations to the two planes which bisect the angles between the planes.

$$3x - 4y + 5z = 3 : 5x + 3y - 4z = 9$$

Also point out which plane bisects the acute angle.

Solution: Given planes are

$$3x - 4y + 5z = 3 \qquad (1)$$

and
$$5x + 3y - 4z = 9 \qquad (2)$$

The equations of the planes bisecting the angles between the given planes are

$$\frac{3x - 4y + 5z - 3}{\sqrt{3^2 + 4^2 + 5^2}} = \pm \frac{5x + 3y - 4z - 9}{\sqrt{5^2 + 3^2 + 4^2}}$$

$$\therefore \qquad 2x + 7y - 9z - 6 = 0 \qquad (3)$$

and
$$8x - y + z - 12 = 0 \qquad (4)$$

which are the required planes.

Let θ be the angle between the plane (3) and either of the given planes

say,
$$5x + 3y - 4z = 9$$

$$\therefore \qquad \cos\theta = \frac{2 \times 5 + 7 \times 3 + (-9)(-4)}{\sqrt{2^2 + 7^2 + 9^2}\ \sqrt{5^2 + 3^2 + 4^2}}$$

$$= \frac{67}{5\sqrt{268}}$$

$$\therefore \qquad \theta = 35° < 45°$$

Now, θ is half the angle between the given planes so that (3) bisects that angle between the planes which is $< 90°$.

Hence, the plane $2x + 7y - 9z = 6$ bisects the acute angle.

(3) Find the equations of the planes bisecting the angle between the planes $x + 2y + 2z = 9$ and $4x - 3y + 12z = 12 = 0$.

Solution: The given planes are

$$x + 2y + 2z = 9 \qquad (1)$$

and
$$4x - 3y + 12z + 12 = 0 \qquad (2)$$

The equations of the planes bisecting the angles between the given planes are:

$$\frac{x + 2y + 2z - 9}{\sqrt{1^2 + 2^2 + 2^2}} = \pm \frac{4x - 3y + 12z + 12}{\sqrt{4^2 + 3^2 + 12^2}}$$

$$\frac{x + 2y + 2z - 9}{3} = \pm \frac{4x - 3y + 12z + 12}{13}$$

$$13x + 26y + 2z - 117 = \pm 12x - 9y + 36z + 36$$

$$\therefore \qquad x + 35y - 34z - 153 = 0 \qquad\qquad (3)$$

$$25x + 17y + 38z - 81 = 0 \qquad\qquad (4)$$

which are the required planes.

Definition: The equation of any plane through the line of intersection of the planes

$$ax + by + cz + d = 0, a_1x + b_1y + c_1z + d_1 = 0 \text{ is}$$

$$(ax + by + cz + d) + \lambda(a_1x + b_1y + c_1z + d_1) = 0 .$$

(4) Obtain the equation of a plane passing through the line of intersection of the planes $7x - 4y + 7z + 16 = 0$ and $4x + 3y - 2z + 13 = 0$ and perpendicular to the plane $x - y - 2z + 5 = 0$

Solution: The equation of any plane through the line of intersection of the two given planes is

$$7x - 4y + 7z + 16 + \lambda(4x + 3y - 2z + 13) = 0$$

$$\text{or} \qquad (7 + 4\lambda)x + (-4 + 3\lambda)y + (7 - 2\lambda)z + 16 + 13\lambda = 0 \qquad\qquad (1)$$

$\therefore$ This plane will be $\perp$ to the plane

$$x - y - 2z + 5 = 0$$

$\therefore$ Their normals are $\perp$

$$\therefore \qquad aa_1 + bb_1 + cc_1 = 0$$

$$(7 + 4\lambda)(1) + (-4 + 3\lambda)(-1) + (7 - 2\lambda)(-2) = 0$$

$$\therefore \qquad \lambda = \frac{3}{5}$$

Substituting in (1)

$$\left(7 + \frac{12}{5}\right)x + \left(-4 + \frac{9}{5}\right)4 + \left(7 - \frac{6}{5}\right)z + \left(16 + \frac{39}{5}\right) = 0$$

$$\therefore \qquad 47x - 11y + 29z + 119 = 0$$

is the required plane.

(5) Find the equation of the plane through the line of intersection of the planes $x + y + z = 1$ and $2x + 3y + 4z = 5$ and perpendicular to the plane $x - y + z = 0$.

Solution: The equation of any plane through the line of intersection of the two given planes is

$$(x + y + z - 1) + \lambda(2x + 3y + 4z - 5) = 0$$

$$\therefore \qquad (1+2\lambda)x + (1+3\lambda)y + (1+4\lambda)z + 1 - 5\lambda = 0 \qquad (1)$$

This plane is $\perp$ to the plane

$$x + y + z = 0$$

$$\therefore \qquad (1+2\lambda)(1) + (1+3\lambda)(-1) + (1+4\lambda)(1) = 0 \quad (\because aa_1 + bb_1 + cc_1 = 0)$$

$$(2-3+4)\lambda + (1-1+1) = 0$$

$$3\lambda + 1 = 0 \quad \therefore \lambda = -\frac{1}{3}$$

Substituting in (1)

$$(x+y+z-1) - \frac{1}{3}(2x+3y+4z-5 = 0$$

$$3x+3y+3z-3-2x-3y-4z+5 = 0$$

$x - z + 2 = 0$ is the required equation of the plane.

(6) Find the equation of the plane through the intersection of the planes $x + y + z = 1$, $2x + 3y + 4z = 7$ and $\perp$ to the plane $x - 5y + 3z = 5$.

Solution: (as in Example 5)

EQUATIONS OF A STRAIGHT LINE

(i) General Form

Two equations of first degree in x, y, z i.e.,

$$ax + by + cz + d = 0 \qquad (1)$$

and $\qquad a_1x + b_1y + c_1z + d_1 = 0 \qquad (2)$

taken together represent a straight line which is the line of intersection of the planes (1) and (2). Thus, the general equations of a straight line are

$$ax + by + cz + d = 0$$

$$a_1x + b_1y + c_1z + d_1 = 0$$

The equations of the x-axis are $y = 0$ and $z = 0$. For, it is the line of intersection of the zx-plane ($y = 0$) and the xy-plane ($z = 0$)

Similarly, the other two.

(ii) Symmetrical Form

Symmetrical form equations of the line passing through the point $P(x_1, y_1, z_1)$ and having direction cosines l, m, n are

Proof:
$$\frac{x - x_1}{l} = \frac{y - y_1}{m} = \frac{z - z_1}{n}$$

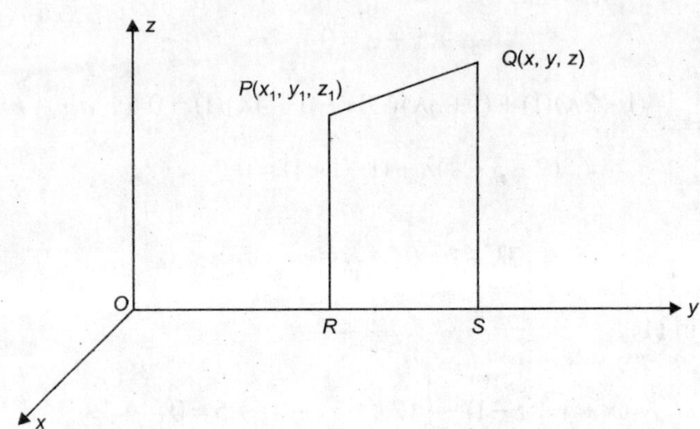

Let $Q(x, y, z)$ be any point on the given line, such that $PQ = r$. Draw PR and $QS \perp$ to the y-axis.

Then
$$RS = OS - OR = y - y_1$$

Also RS = projection of PQ on the y-axis.
$$= r \cos \beta = rm.$$

where β is the angle PQ makes with OY.

$\therefore y - y_1 = mr$, similarly, $z - z_1 = nr$, $x - x_1 = lr$.

Hence, the required equations are

$$\frac{x - x_1}{l} = \frac{y - y_1}{m} = \frac{z - z_1}{n} (= r) \tag{1}$$

Cor 1: Any point on the line (1) is

$$(x_1 + lr,\ y_1 + mr,\ z_1 + nr)$$

Cor 2: The equations of the line through (x_1, y_1, z_1) and having direction cosines proportional to a, b, c are

$$\frac{x - x_1}{a} = \frac{y - y_1}{b} = \frac{z - z_1}{c}$$

Cor 3: The equations of the line joining the points (x_1, y_1, z_1) and (x_2, y_2, z_2) are

$$\frac{x - x_1}{x_2 - x_1} = \frac{y - y_1}{y_2 - y_1} = \frac{z - z_1}{z_2 - z_1}$$

where $x_2 - x_1, y_2 - y_1, z_2 - z_1$, are the direction ratios.

Examples

(1) Find the distance of the point (1, –2, 3) from the plane $x - y + z = 5$ measured parallel to the line $\dfrac{x}{2} = \dfrac{y}{3} = \dfrac{z}{-6}$.

Solution: The line through $P(1, -2, 3)$ having direction cosines 2, 3, –6 is

$$\frac{x-1}{2} = \frac{y+2}{3} = \frac{z-3}{-6} = r$$

Any point will be on the plane is

$$(2r+1, 3r-2, 3-6r)$$

$$\therefore \qquad x - y + 2 = 5$$

$$2r+1-(3r-2)+3-6r = 5 \qquad\qquad \therefore r = \frac{1}{7}$$

$\therefore$ The point of intersection is $Q\left(\dfrac{9}{7}, \dfrac{-11}{7}, \dfrac{15}{7}\right)$

$\therefore$ The required distance PQ

$$= \sqrt{\frac{4}{49} + \frac{8}{49} + \frac{36}{49}}$$

$$= \sqrt{\frac{4+9+36}{49}} = \sqrt{\frac{49}{49}}$$

$$= 1$$

(2) Find the equations of the straight line in the symmetrical form of the line

$$x + y + z + 1 = 0 \text{ and } 4x + y - 2z + 2 = 0.$$

Solution: (i) To find a point on the line.

Put $z = 0$ in the given equations, we get

$$\left.\begin{array}{r} x + y + 1 = 0 \\ 4x + y + 2 = 0 \end{array}\right\}$$

Solving

$$x = -\frac{1}{3}, \; y = -\frac{2}{3} \text{ and } z = 0$$

$\therefore$ A point on the line is $\left(-\dfrac{1}{3}, \dfrac{-2}{3}, 0\right)$

(ii) To find the direction cosines l, m, n of a line. Since the line lies on both the given planes, it is $\perp$ to their normals whose direction cosines are proportional to 1, 1, 1 and 4, 1, –2

$\therefore$ $l + m + n = 0$

$\qquad 4l + m - 2n = 0 \qquad (\because ll_1 + mm_1 + nn_1 = 0)$

Solving by cross multiplication, we get

$$\frac{l}{-1} = \frac{m}{2} = \frac{n}{-1}$$

$\therefore$ The direction cosines of the given line are proportional to –1, 2, –1.

Hence, the equation of the straight line in the symmetrical form is

$$\frac{x + \dfrac{1}{3}}{-1} = \frac{y + \dfrac{2}{3}}{2} = \frac{z - 0}{-1}.$$

(3) Find the equation of the plane passing through the line of intersection of the planes $2x + y - z = 1$; $5x - 3y + 4z + 3 = 0$ and parallel to the line

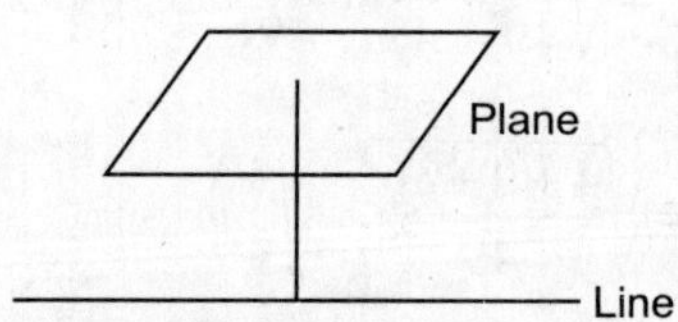

$$\frac{x - 1}{2} = \frac{y - 2}{3} = \frac{z - 3}{4}$$

Solution: The equation of any plane through the intersection of the given planes is

$$(2x + y - z - 1) + \lambda(5x - 3y + 4z + 3) = 0$$

or $\qquad (2 + 5\lambda)x + (1 - 3\lambda)y + (4\lambda - 1)z + 3\lambda - 1 = 0 \qquad\qquad (1)$

Hence, the direction ratios of the normal to this plane are

$$5\lambda + 2, \ 1 - 3\lambda, \ 4\lambda - 1.$$

Direction ratios of the given line are

$$2, 3, 4.$$

Since the plane (1) is parallel to the given line, normals to the line should be perpendicular to the given line.

$$\therefore \quad aa' + bb' + cc' = 0$$

$$2(5\lambda + 2) + 3(1 - 3\lambda) + 4(4\lambda - 1) = 0$$

$$\therefore \qquad k = \frac{-3}{17} .$$

Hence, the required equation of the plane is

$$(2x + y - z - 1) - \frac{3}{17}(5x - 3y + 4z + 3) = 0$$

i.e., $\qquad 19x + 26y - 29z - 26 = 0$

(4) Find the equations of the line passing through the point $A(4, 2, -2)$ and $\perp$ to lines $\dfrac{x+1}{-1} = \dfrac{y+2}{2} = \dfrac{z-3}{3}$ and $\dfrac{x-2}{3} = \dfrac{y}{4} = \dfrac{z-1}{2}$.

Solution: Any line passing through $A(4, 2, -2)$ is of the form

$$\frac{x-4}{a} = \frac{y-2}{b} = \frac{z+2}{c} \tag{1}$$

where a, b, c are direction ratios.

Now, the direction ratios of the given lines are

$$-1, 2, 3 \text{ and } 3, 4, 2$$

Since the line (1) is $\perp$ to the given lines

$$-a + 2b + 3c = 0$$

$$3a + 4b + 2c = 0$$

Solving by cross multiplication, we get

$$\frac{a}{-8} = \frac{b}{11} = \frac{c}{-10}$$

$\therefore$ The required equations of the line are

$$\frac{x-4}{-8} = \frac{4-2}{11} = \frac{2+2}{-10} .$$

(5) Find the equation of the plane through the intersection of the planes $2x - y + z = 4$ and $3x + 4y - z + 5 = 0$ and the point $(1, 1, 1)$.

Solution: The equation of any plane through the intersection of the given planes is

$$(2x - y + z - 4) + \lambda(3x + 4y - z + 5) = 0 \tag{1}$$

This plane passes through the point $(1, 1, 1)$

$$\therefore \qquad (2 - 1 + 1 - 4) + \lambda(3 + 4 - 1 + 5) = 0$$

$$-2 + 11\lambda = 0 \quad \therefore \lambda = \frac{2}{11}$$

$$\therefore \qquad (2x - y + z - 4) + \frac{2}{11}(3x + 4y - z + 5) = 0$$

$$(22x - 11y + 11z - 44) + (6x + 8y - 2z + 10) = 0$$

$$\therefore \qquad 16x - 3y + 9z - 33 = 0$$

is the required equation of the plane.

ANGLE BETWEEN THE LINE $\dfrac{x - x_1}{l} = \dfrac{y - y_1}{m} = \dfrac{z - z_1}{n}$ AND THE PLANE $ax + by + cz + d = 0$

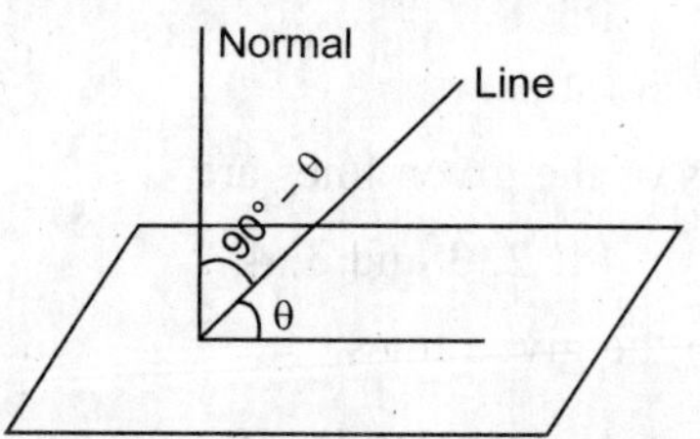

Let θ be the angle between the line and the plane then $90° - \theta$ is the angle between the line and the normal to the plane.

Now, the direction ratios of the line are l, m, n and the direction ratios of the normal to the plane are a, b, c.

$$\therefore \qquad \cos(90° - \theta) = \frac{la - mb + nc}{\sqrt{l^2 + m^2 + n^2}\,\sqrt{a^2 + b^2 + c^2}}$$

$$\therefore \qquad \sin\theta = \frac{la + mb + nc}{\sqrt{\Sigma l^2}\,\sqrt{\Sigma a^2}}$$

$$\therefore \qquad \theta = \sin^{-1}\left(\frac{la + mb + nc}{\sqrt{\Sigma l^2}\,\sqrt{\Sigma a^2}}\right)$$

Cor 1: If the line is parallel to the plane $\sin \theta = 0$

$\therefore \qquad\qquad\qquad al + bm + cn = 0$

Cor 2: If the line is perpendicular to the plane then

$$\frac{l}{a} = \frac{m}{b} = \frac{n}{c}.$$

Conditions for a line to lie in a plane:

The conditions that the line

$$\frac{x - x_1}{l} = \frac{y - y_1}{m} = \frac{z - z_1}{n} \tag{1}$$

may lie in the plane

$$ax + by + cz + d = 0 \tag{2}$$

$\text{are} \qquad\qquad al + bm + cn = 0$

$\text{and} \qquad\qquad ax_1 + by_1 + cz_1 + d = 0$

And the equation of any plane through the line

$$\frac{x - x_1}{l} = \frac{y - y_1}{m} = \frac{z - z_1}{n}$$

$\text{is} \qquad a(x - x_1) + b((y - y_1) + c(z - z_1) = 0$

Condition for the two lines intersect (or to be coplanar)

Let the equations of the lines be

$$\frac{x - x_1}{l_1} = \frac{y - y_1}{m_1} = \frac{z - z_1}{n_1} \tag{1}$$

$\text{and} \qquad \dfrac{x - x_2}{l_2} = \dfrac{y - y_2}{m_2} = \dfrac{z - z_1}{n_2} \qquad\qquad (2)$

The equation of any plane through the line (1) is

$$a(x - x_1) + b(y - y_1) + c(z - z_1) = 0 \tag{3}$$

$\text{where} \qquad\qquad al_1 + bm_1 + cn_1 = 0 \qquad\qquad (4)$

The line (2) will lie in the plane (3) if it is parallel to the plane and its point (x_2, y_2, z_2) lie on this plane.

$\therefore \qquad\qquad al_2 + bm_2 + cn_2 = 0 \qquad\qquad (5)$

and
$$a(x_2 - x_1) + b(y_2 - y_1) + c(z_2 - z_1) = 0 \tag{6}$$

Eliminating a, b, c from (6), (5) and (4), we get

$$\begin{vmatrix} x_2 - x_1 & y_2 - y_1 & z_2 - z_1 \\ l_1 & m_1 & n_1 \\ l_2 & m_2 & n_2 \end{vmatrix} = 0$$

is the required condition.

Also eliminating a, b, c from (3), (4) and (5), we get

$$\begin{vmatrix} x - x_1 & y - y_1 & z - z_1 \\ l_1 & m_1 & n_1 \\ l_2 & m_2 & n_2 \end{vmatrix} = 0$$

which is the equation of the plane containing the lines (1) and (2)

Examples

(1) Show that the lines

$$\frac{x-5}{4} = \frac{y-7}{4} = \frac{z+3}{-5} \text{ and } \frac{x-8}{7} = \frac{y-4}{1} = \frac{z-5}{3}$$

are co-planar. Find their common point and the equation of the plane in which they lie.

Solution: Let the given equations be

$$\frac{x-5}{4} = \frac{y-7}{4} = \frac{z+3}{-5} \tag{1}$$

and
$$\frac{x-8}{7} = \frac{y-4}{1} = \frac{z-5}{3} \tag{2}$$

Any point on the line (1) is

$$(5 + 4r, \, 7 + 4r, \, -3 - 5r) \tag{3}$$

Point (3) lies on (2) if

$$\frac{-3 + 4r}{7} = \frac{3 + 4r}{1} = \frac{-8 - 5r}{3} \tag{4}$$

$$\therefore \quad \frac{-3 + 4r}{7} = 3 + 4r \Rightarrow r = -1$$

This value clearly satisfies

$$\frac{3+4r}{1}=\frac{-8-5r}{3} \qquad \therefore r=-1$$

This proves that the lines are coplanar.

Substituting in (3) the value of r, we get (1, 3, 2) which is the common point. The equation of the plane in which they lie is

$$\begin{vmatrix} x-5 & y-7 & z+3 \\ 4 & 4 & -5 \\ 7 & 1 & 3 \end{vmatrix}=0$$

i.e., $17x-47y-24z+172=0$

(2) Find the equation of the plane through (2, –1, 6) (1, –2, 4) and perpendicular to the plane $x-2y-2z+9=0$.

Solution: Equation of the plane through (2, –1, 6) is

$$a(x-2)+b(y+1)+c(z-6)=0 \tag{1}$$

It passes through (1, –2, 4)

$$\therefore \qquad a(1-2)+b(-2+1)+c(4-6)=0$$

$$\therefore \qquad -a+b-2c=0$$

$$\text{or} \qquad a+b+2c=0 \tag{2}$$

By the condition of perpendicularity (1) is perpendicular to $x-2y-2z+9=0$

$$\therefore \qquad a(1)+b(-2)+c(-2)=0$$

$$\therefore \qquad a-2b-2c=0 \tag{3}$$

Eliminate a, b, c from equations (1) (2) & (3), we get

$$\begin{vmatrix} x-2 & y+1 & z-6 \\ 1 & 1 & 2 \\ 1 & -2 & -2 \end{vmatrix}=0$$

$\therefore \quad 2x+4y-3z+18=0$ which is the equation of the required plane.

(3) Find the equations of the straight line perpendicular to the lines

$$\frac{x-3}{2}=\frac{y-2}{4}=\frac{z-4}{3}; \text{ and } \frac{x-4}{1}=\frac{y-3}{3}=\frac{z-3}{4}$$

and passing through their points of intersection.

Solution: Any point on the first line is

$$P(2r+3,\ 4r+2,\ 3r+4)$$

Since this point also lies on the second line, we get

$$\frac{2r+3-4}{1} = \frac{4r+2-3}{3} = \frac{3r+4-3}{4}$$

$$\therefore \quad 2r-1 = \frac{4r-1}{3} = \frac{3r+1}{4}$$

$$\therefore \quad 2r-1 = \frac{4r-1}{3} \Rightarrow r = -1$$

$\therefore$　point of intersection is $P = (5,\ 6,\ 7)$.

Any line through $P(5,\ 6,\ 7)$ is

$$\frac{x-5}{a} = \frac{y-6}{b} = \frac{z-7}{c}$$

Since this line is $\perp$ to both the given lines, we have

$$2a+4b+3c = 0$$

$$a+3b+4c = 0$$

Solving,　$\dfrac{a}{7} = \dfrac{b}{-5} = \dfrac{c}{2}$

$\therefore$ The required line is

$$\frac{x-5}{7} = \frac{y-6}{-5} = \frac{z-7}{2}.$$

SHORTEST DISTANCE BETWEEN TWO LINES

Definition: Two straight lines which do not lie in one plane are called skew lines, such lines possess a common perpendicular which is the shortest distance between them.

Let the given skew lines AB and CD be

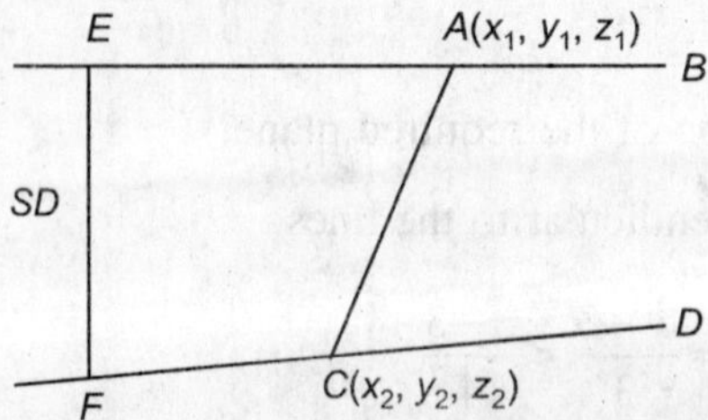

$$\frac{x-x_1}{l_1} = \frac{y-y_1}{m_1} = \frac{z-z_1}{n_1}$$

and
$$\frac{x - x_2}{l_2} = \frac{y - y_2}{m_2} = \frac{z - z_2}{n_2}$$

where $A = (x_1, y_1, z_1)$ and $C = (x_2, y_2, z_2)$

Let l, m, n be the direction cosines of the shortest distance EF. Since EF. is $\perp$ to both AB and CD.

$\therefore$
$$ll_1 + mm_1 + nn_1 = 0$$

$$ll_2 + mm_2 + nn_2 = 0$$

Solving,
$$\frac{l}{m_1 n_2 - m_2 n_1} = \frac{m}{n_1 l_2 - n_2 l_1} = \frac{n}{l_1 m_2 - m_1 l_2}$$

$$= \frac{\sqrt{l^2 + m^2 + n^2}}{\sqrt{\Sigma(m_1 n_2 - m_2 n_1)^2}} = \frac{1}{\sin\theta} \tag{1}$$

where θ is the angle between the lines AB and CD.

$\therefore$ length of $SD(EF)$ = Projection of AC on EF

$$= l(x_2 - x_1) + m(y_2 - y_1) + n(z_2 - z_1).$$

where l_1, m_1, n_1 are given by (1)

To find the equation of the line of shortest distance. We observe that it is coplanar with both AB and CD. Plane containing the line AB and EF is

$$\begin{vmatrix} x - x_1 & y - y_1 & z - z_1 \\ l_1 & m_1 & n_1 \\ l & m & n \end{vmatrix} = 0 \tag{2}$$

and the plane containing line CD and EF is

$$\begin{vmatrix} x - x_2 & y - y_2 & z - z_2 \\ l_2 & m_2 & n_2 \\ l & m & n \end{vmatrix} = 0 \tag{3}$$

Hence, (2) and (3) are the equations of the line of shortest distance.

Examples

(1) Find the length and the equations of the shortest distance between the lines.

$$\frac{x-1}{2} = \frac{y-2}{3} = \frac{z-3}{4} \text{ and } \frac{x-2}{3} = \frac{y-4}{4} = \frac{z-5}{5}$$

Solution: Let the given two lines be

$$\frac{x-1}{2} = \frac{y-2}{3} = \frac{z-3}{4} \qquad (1)$$

$$\frac{x-2}{3} = \frac{y-4}{4} = \frac{z-5}{5} \qquad (2)$$

Let l, m, n be the direction cosines of the shortest distance EF since it is $\perp$ to both the lines AB and CD, we have

$$2l + 3m + 4n = 0$$

$$3l + 4m + 5n = 0$$

Solving, $\dfrac{1}{-1} = \dfrac{m}{2} = \dfrac{n}{-1} = \dfrac{1}{\sqrt{6}}$ so that

$$l = \frac{-1}{\sqrt{6}}, \; m = \frac{2}{\sqrt{6}}, \; n = \frac{-1}{\sqrt{6}}$$

Hence, the $SD(EF)$ is

$$= \text{Projection of } AC \text{ on } EF$$

$$= l(x_2 - x_1) + m(y_2 - y_1 + n(z_2 - z_1)$$

$$= \frac{1}{\sqrt{6}}(2-1) + \frac{2}{\sqrt{6}}(4-2) - \frac{1}{6}(5-3) = \frac{1}{\sqrt{6}}.$$

Equations of the line of shortest distance are

$$\begin{vmatrix} x-1 & y-2 & z-3 \\ 2 & 3 & 4 \\ -1 & 2 & -1 \end{vmatrix} = 0 \text{ and } \begin{vmatrix} x-2 & y-4 & z-5 \\ 3 & 4 & 5 \\ -1 & 2 & -1 \end{vmatrix} = 0$$

$$\therefore \; 11x + 2y - 7z + 6 = 0 \text{ and } 7x + 4 - 5z + 7 = 0.$$

(2) Find the magnitude and the equation of the shortest distance between the lines

$$\frac{x}{2} = \frac{y}{-3} = \frac{z}{1}; \; \frac{x-2}{3} = \frac{y-1}{-5} = \frac{z+2}{2}$$

Solution: Let l, m, n be the direction cosines of the shortest distance EF. Since EF is $\perp$ to both AB and CD, we have

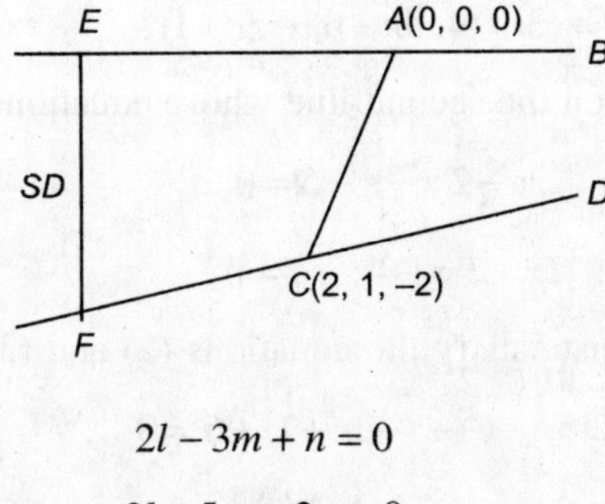

$$2l - 3m + n = 0$$

$$3l - 5m + 2n = 0$$

Solving by cross multiplication

$$\frac{l}{1} = \frac{m}{1} = \frac{n}{1} = \frac{\sqrt{l^2 + m^2 + n^2}}{\sqrt{1^2 + 1^2 + 1^2}} = \frac{1}{\sqrt{3}}$$

$\therefore$
$$l = \frac{1}{\sqrt{3}}, \ m = \frac{1}{\sqrt{3}}, \ n = \frac{1}{\sqrt{3}}.$$

$\therefore$ Length of $SD(EF)$ = Projection of AC on EF.

$$= l(x_2 - x_1) + m(y_2 - y_1) + n(z_2 - z_1)$$

$$= \frac{1}{\sqrt{3}}(2 - 0) + \frac{1}{\sqrt{3}}(1 - 0) + \frac{1}{\sqrt{3}}(-2 - 0)$$

$$= \frac{1}{\sqrt{3}}.$$

Equation of the line $SD(EF)$ are

$$\begin{vmatrix} x & y & z \\ 2 & -3 & 1 \\ 1 & 1 & 1 \end{vmatrix} = 0 \text{ and } \begin{vmatrix} x-2 & y-1 & z+2 \\ 3 & -5 & 2 \\ 1 & 1 & 1 \end{vmatrix} = 0$$

$\therefore \quad 4x + y + 5z = 0$ and $7x + y - 8z = 31$.

(3) Show that the lines $\dfrac{x+4}{3} = \dfrac{y+6}{5} = \dfrac{z-1}{-2}$ and $3x - 2y + z + 5 = 0$;

$2x + 3y + 4z - 4 = 0$ are coplanar. Find their point of intersection and the plane containing them.

Solution: Any point on the first line is

$$P \equiv (3r - 4, 5r - 6, -zr + 1) \tag{1}$$

Since this point also lies on the second line whose equations are

$$3x - 2y - z - 5 = 0 \tag{2}$$

$$2x + 3y + 4z - 4 = 0 \tag{3}$$

The co-ordinates of P must satisfy the equations (2) and (3)

$$\therefore \qquad 3(3r - 4) - 2(5r - 6) + (-2r + 1) + 5 = 0 \quad \therefore r = 2$$

and $\qquad 2(3r - 4) + 3(5r - 6) + 4(-2r + 1) - 4 = 0 \quad \therefore r = 2$

since the two values of r are equal, the given lines intersect and their point of intersection is $P(2, 4, -3)$.

The equation of the plane containing the second line is

$$(3x - 2y + z + 5) + \lambda(2x + 3y + 4z - 4) = 0 \tag{4}$$

This will contain the first line, if its point $(-4, -6, 1)$ lies on it.

$$\therefore \qquad -12 + 12 + 1 + 5\,k\,(-8 - 18 + 4 - 4) = 0$$

$$\therefore \qquad k = \frac{3}{13}$$

Substituting the value of k in (4), we get

$$45x - 17y + 25z + 33 = 0$$

QUESTION BANK 2

(1) Find the equation of the plane passing through the points $(3, 1, 2)$ and $(3, 4, 4)$ and perpendicular to $5x + y + 4z = 0$.

(2) Show that the points $(2, 2, 0)$ $(4, 5, 1)$ $(3, 9, 4)$ and $(0, -1, -1)$ are co-planar. Find the equation of the plane containing them.

(3) Find the equation of a straight line through $(7, 2, -3)$ and perpendicular to each of the lines

$$\frac{x - 2}{3} = \frac{y - 3}{4} = \frac{z - 4}{5} \text{ and } \frac{z + 2}{4} = \frac{y - 3}{5} = \frac{z - 4}{6}.$$

(4) Derive the equation to the plane in the intercept form $\dfrac{x}{a} + \dfrac{y}{b} + \dfrac{z}{c} = 1$.

(5) Find the equation of the plane which passes through the point $(3, -3, 1)$ and is normal to the line joining the points $(3, 2, -1)$ and $(2, -1, 5)$.

(6) Find the angle between the planes $x - y + z - 6 = 0$ and $2x + 3y + z + 5 = 0$.

(7) Find the equation of the plane through the line of intersection of the planes $x + y + z = 1$ and $2x + 3y + 4z = 5$ and perpendicular to the plane $x - y + z = 0$.

(8) Find the image of the point $(1, -2, 3)$ in the plane $2x + y - z = 5$.

(9) Show that the lines $\dfrac{x-5}{4} = \dfrac{y-7}{4} = \dfrac{z+3}{-5}$ and $\dfrac{x-8}{7} = \dfrac{y-4}{1} = \dfrac{z-5}{3}$ intersect. Find their point of intersection and the equation of the plane in which they lie.

(10) With usual notations, derive the equation of the plane in the form $lx + my + nz = p$.

(11) Find the angle betwen the planes $2x + 4y - 6z = 1$ and $3x + 6y + 5z + 4 = 0$.

(12) Find the distance of the point $(1, -2, 3)$ from the plane $x - y + z = 5$ measured parallel to the line $\dfrac{x}{2} = \dfrac{y}{3} = \dfrac{z}{-6}$.

(13) Obtain the equation of the plane passing through the line of intersection of the planes $7x - 4y + 7z + 16 = 0$ and $4x + 3y - 2z + 13 = 0$ and perpendicular to the plane $x - y - 2z + 5 = 0$.

(14) Find the angle between the planes $2x - y + z = 1$ and $x + y + 2z = 3$.

(15) Find the image of the point $A(2, -1, 0)$ in the line $\dfrac{x-2}{2} = \dfrac{y-1}{1} = \dfrac{z-3}{1}$.

PART (B)
Vector Algebra and Vector Calculus

UNIT 3

Vector Algebra

SCALARS AND VECTORS

A quantity which is completely specified by its magnitude only is called a scalar. For example, mass, time, length, temperature, work, electric charge, statistical data are all scalar quantities.

A quantity which is completely specified by its magnitude as well as direction is called a vector. For example, weight, displacement, velocity, acceleration, force are all vector quantities.

A vector is represented by a directed line segment. Let P be the initial point and Q be the terminal point. Then the line joining PQ is a vector and is denoted by $\overrightarrow{PQ}$ and its magnitude is denoted by $|\overrightarrow{PQ}| = PQ$.

Generally, vectors are denoted by bold letters or by using arrow marks on the top, like $\overrightarrow{PQ}, \overrightarrow{AB}$, etc.

TYPES OF VECTORS

(1) Unit vector: A vector whose magnitude is unity is called a unit vector. If $\vec{a}$ is any vector, then $\hat{a}$ is a unit vector in the direction of $\overline{a}$. Then $\overline{a} = |\overline{a}| \hat{a} \Rightarrow \hat{a} = \dfrac{\overline{a}}{|\overline{a}|}$.

(2) Null or zero vector: A vector whose magnitude is zero is called a null or zero vector and is denoted by $\overline{0}$. A non-zero vector is called a proper vector.

(3) Co-initial vectors: Two or more vectors having the same initial point are called co-initial vectors.

(4) Equal vectors: Two vectors are said to be equal if they have the same magnitude and the same or parallel directions.

(5) Like and unlike vectors: Two vectors are said to be parallel if they lie along the same or parallel lines irrespective of their directions. However, if their sense is also the same, they are called like parallel vectors and if their sense is opposite, they are called unlike parallel vectors.

(6) Position vector of a point: The position vector of a point P with respect to an arbitrary chosen origin O is the vector $\overrightarrow{OP}$. Generally, the position vector we take

$$\overrightarrow{OP} = \vec{R} = x\hat{i} + y\hat{j} + z\hat{k}.$$

ADDITION OF VECTORS

Let $\vec{a}, \vec{b}$ be two given vectors represented by $\overrightarrow{OA}$ and $\overrightarrow{OB}$ respectively (so that

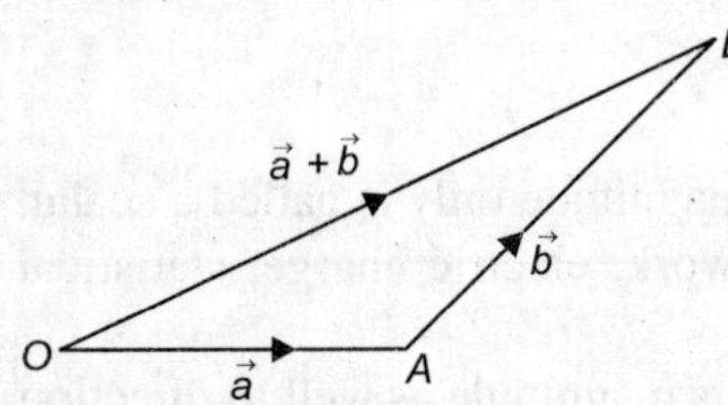

the terminal point of $\vec{a}$ is the initial point of $\vec{b}$). Then the vector $\overrightarrow{OP}$ represents the sum (or resultant) of $\vec{a}$ and $\vec{b}$ and is written as $\vec{a} + \vec{b}$.

Thus, $\overrightarrow{OA} + \overrightarrow{AB} = \vec{a} + \vec{b} = \overrightarrow{OB}$

Properties

(1) Vector addition is commutative

i.e., $\qquad\qquad \vec{a} + \vec{b} = \vec{b} + \vec{a}$

(2) Vector addition is associative

i.e., $\qquad\qquad \vec{a} + (\vec{b} + \vec{c}) = (\vec{a} + \vec{b}) + \vec{c}$

(3) For any vector $\vec{a}$, $\vec{a} + \vec{0} = \vec{0} + \vec{a} = \vec{a}$

(4) For any vector $\vec{a}$, $\vec{a} + (-\vec{a}) = \vec{0}$

(5) Multiplication of a vector by a scalar.

Let $\vec{a}$ be a vector and m be a scalar, then their product $m\vec{a}$ is a vector whose magnitude is $|m|$ times that of $\vec{a}$ and the direction is the same or opposite to $\vec{a}$ as m is positive or negative.

If $\vec{a}$ and $\vec{b}$ are two vectors and m, n are scalars, then

(a) $m(n\vec{a}) = n(m\vec{a}) = (mn)\vec{a}$

(b) $(m + n)\vec{a} = m\vec{a} + n\vec{a}$

(c) $m(\vec{a} + \vec{b}) = m\vec{a} + m\vec{b}$.

(6) Collinear vectors: If $\vec{a}$ and $\vec{b}$ are collinear vectors, then $\vec{a} = x\vec{b}$ where x is a scalar.

SCALAR (OR DOT) PRODUCT OF TWO VECTORS

Definition

The scalar product of two vectors $\vec{a}$ and $\vec{b}$ is denoted by $\vec{a} \cdot \vec{b}$ (read as $\vec{a}$ dot $\vec{b}$) and is defined as

$$\vec{a} \cdot \vec{b} = |\vec{a}||\vec{b}|\cos\theta$$

$$= ab\cos\theta$$

where θ is the angle between $\vec{a}$ and $\vec{b}$.

Since $|\vec{a}|, |\vec{b}|, \cos\theta$ are scalars, hence $\vec{a}.\vec{b}$ is a scalar.

Properties

(1) Scalar product of two vectors is commutative

i.e., $\qquad \vec{a} \cdot \vec{b} = \vec{b} \cdot \vec{a}$, each $= ab \cos \theta$.

(2) Two non-zero vectors are perpendicular if and only if their scalar product is zero.

If $\vec{a}$ and $\vec{b}$ are perpendicular, then

$$\vec{a} \cdot \vec{b} = ab\cos 90° = 0$$

$\therefore \vec{a}$ is perpendicular to $\vec{b}$.

(3) If $\hat{i}, \hat{j}, \hat{k}$ are unit vectors along the co-ordinate axes, then

$$\hat{i} \cdot \hat{j} = \hat{j} \cdot \hat{k} = \hat{k} \cdot \hat{i} = 0$$

and $\qquad \hat{i} \cdot \hat{i} = \hat{j} \cdot \hat{j} = \hat{k} \cdot \hat{k} = 1$

(4) Scalar product of vectors is distributive

If $\vec{a}, \vec{b}.\vec{c}$ are three vectors, then

$$\vec{a} \cdot (\vec{b} + \vec{c}) = \vec{a} \cdot \vec{b} + \vec{a} \cdot \vec{c}$$

(5) If $\vec{a} = a_1\hat{i} + a_2\hat{j} + a_3\hat{k}, \vec{b} = b_1\hat{i} + b_2\hat{j} + b_3\hat{k}$, then

$$\vec{a} \cdot \vec{b} = (a_1\hat{i} + a_2\hat{j} + a_3\hat{k}) \cdot (b_1\hat{i} + b_2\hat{j} + b_3\hat{k})$$

$$= a_1b_1 + a_2b_2 + a_3b_3.$$

(6) The angle between two vectors $\vec{a}$ and $\vec{b}$ is

$$\vec{a} \cdot \vec{b} = |\vec{a}||\vec{b}|\cos\theta.$$

$$\therefore \quad \cos\theta = \frac{\vec{a}\cdot\vec{b}}{|\vec{a}||\vec{b}|}$$

$$= \frac{(a_1\hat{i} + a_2\hat{j} + a_3\hat{k}) \cdot (b_1\hat{i} + b_2\hat{j} + b_3\hat{k})}{\sqrt{a_1^2 + a_2^2 + a_3^2}\sqrt{b_1^2 + b_2^2 + b_3^2}}$$

$$\therefore \quad \cos\theta = \frac{a_1b_1 + a_2b_2 + a_3b_3}{\sqrt{a_1^2 + a_2^2 + a_3^2}\sqrt{b_1^2 + b_2^2 + b_3^2}}.$$

(7) Projection of $\vec{a}$ on $\vec{b}$

Projection of $\vec{a}$ on $\vec{b} = (|\vec{a}|\cos\theta)\vec{b}$.

(8) Work done as a scalar product

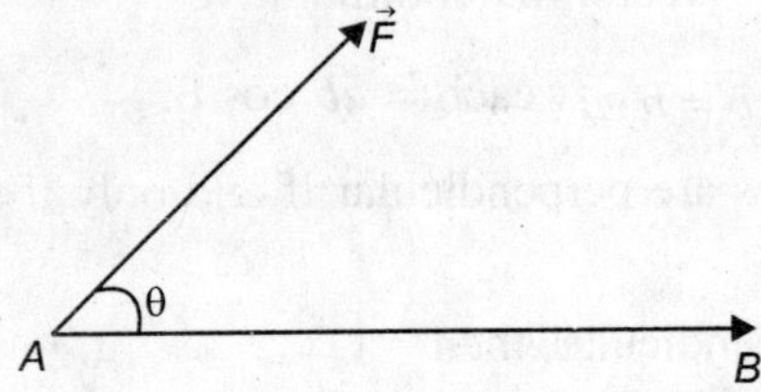

Let a constant force $\vec{F}$ displace a particle from A to B, then the work done by the force is given by

$$W = (\text{Resolved part of } \vec{F} \text{ in the direction of } AB)\, AB$$

$$= (F\cos\theta)AB = |\vec{F}||\overrightarrow{AB}|\cos\theta$$

$$= \vec{F} \cdot \overrightarrow{AB}.$$

Thus, the work done by a constant force is the scalar product of the vectors representing the force and the displacement.

Examples

(1) If $\vec{a} = \hat{i} + 2\hat{j} + 3\hat{k}$, $\vec{b} = -\hat{i} + 2j + \hat{k}$ and $\vec{c} = 3\hat{i} + \hat{j}$, find t such that $\vec{a} + t\vec{b}$ is perpendicular to $\vec{c}$.

Solution: $\vec{a} = \hat{i} + 2\hat{j} + 3\hat{k}$, $\vec{b} = -\hat{i} + 2j + \hat{k}$ and $\vec{c} = 3\hat{i} + \hat{j}$,

Now,

$$\vec{a} + t\vec{b} = (\hat{i} + 2\hat{j} + 3\hat{k}) + t(-\hat{i} + 2\hat{j} + \hat{k})$$

$$= (1-t)\hat{i} + (2+2t)\hat{j} + (3+t)\hat{k}$$

Since $\vec{a} + t\vec{b}$ is perpendicular to $\vec{c}$

$$(\vec{a} + t\vec{b}) \cdot \vec{c} = 0$$

$$\therefore \qquad (1-t)(3) + (2+2t)(1) + (3+t)(0) = 0$$

$$\therefore \qquad t = 5$$

(2) Find the angle and sides of the triangle whose vertices are

$$\hat{i} - 2\hat{j} + 2\hat{k},\ 2\hat{i} + \hat{j} - \hat{k} \text{ and } 3\hat{i} - \hat{j} + 2\hat{k}.$$

Solution: Let $\overrightarrow{OA} = \hat{i} + 2\hat{j} + 2\hat{k}$, $\overrightarrow{OB} = 2\hat{i} + \hat{j} - \hat{k}$ and $\overrightarrow{OC} = 3\hat{i} - \hat{j} + 2\hat{k}$.

Then

$$\overrightarrow{BC} = \overrightarrow{OC} - \overrightarrow{OB} = \hat{i} - 2\hat{j} + 3\hat{k}$$

$$\overrightarrow{CA} = \overrightarrow{OA} - \overrightarrow{OC} = -2\hat{i} - \hat{j}$$

$$\overrightarrow{AB} = \overrightarrow{OB} - \overrightarrow{OA} = \hat{i} + 3\hat{j} - 3\hat{k}$$

$$\therefore \quad BC = |\overrightarrow{BC}| = \sqrt{1+4+9} = \sqrt{14}$$

$$CA = |\overrightarrow{CA}| = \sqrt{4+1} = \sqrt{5}$$

$$AB = |\overrightarrow{AB}| = \sqrt{1+9+9} = \sqrt{19}$$

Now,

$$\therefore \quad \cos A = \frac{\overrightarrow{AB} \cdot \overrightarrow{AC}}{|\overrightarrow{AB}||\overrightarrow{AC}|} = \frac{(\hat{i} + 3\hat{j} - 3\hat{k}) \cdot (2\hat{i} + \hat{j})}{\sqrt{19} \times \sqrt{5}} = \frac{2+3}{\sqrt{19}\sqrt{5}} = \frac{\sqrt{5}}{\sqrt{19}}$$

$$\therefore \quad A = \cos^{-1}\left(\sqrt{\frac{5}{19}}\right)$$

$$\cos B = \frac{\overrightarrow{BA} \cdot \overrightarrow{BC}}{|\overrightarrow{BA}||\overrightarrow{BC}|} = \frac{(-\hat{i} - 3\hat{j} - 3\hat{k}) \cdot (\hat{i} - 2\hat{j} + 3\hat{k})}{\sqrt{19} \times \sqrt{14}} = \frac{-1+6+9}{\sqrt{19}\sqrt{14}} = \frac{14}{\sqrt{19}\sqrt{14}} = \frac{\sqrt{14}}{\sqrt{19}}$$

$$\therefore \quad B = \cos^{-1}\left(\sqrt{\frac{14}{19}}\right)$$

and $\cos C = \dfrac{\overrightarrow{CA} \cdot \overrightarrow{CB}}{|\overrightarrow{CA}||\overrightarrow{CB}|} = \dfrac{(-2\hat{i} - \hat{j}) \cdot (-\hat{i} + 2\hat{j} - 3\hat{k})}{\sqrt{5} \times \sqrt{14}} = \dfrac{2 - 2}{\sqrt{5} \times \sqrt{14}} = 0$

$$C = \cos^{-1}(0) = 90°$$

(3) Determine the value of a so that the vectors

$$\vec{A} = 2\hat{i} + a\hat{j} + \hat{k} \text{ and } \vec{B} = 4\hat{i} - 2\hat{j} - 2\hat{k} \text{ are perpendicular.}$$

Solution: The vectors $\vec{A}$ and $\vec{B}$ are perpendicular if

$$\vec{A} \cdot \vec{B} = 0$$

$\therefore \quad (2\hat{i} + a\hat{j} + \hat{k}) \cdot (4\hat{i} - 2\hat{j} - 2\hat{k}) = 0$

$$8 - 2a - 2 = 0 \Rightarrow a = 3$$

(4) Find the angle between any two diagonals of a cube.

Solution: Consider a cube whose sides are unity.

Let the unit vectors $\hat{i}, \hat{j}, \hat{k}$ be represented by the co-terminous edges

$\overrightarrow{OA}, \overrightarrow{OB}$ and $\overrightarrow{OC}$ respectively.

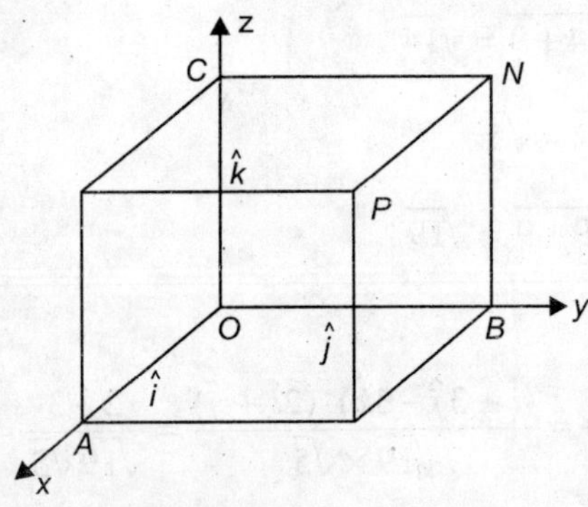

The angle between any two diagonals of a cube is the same.

Let us find the angle between the diagonals OP and AN.

$$\overrightarrow{OP} = ON + NP = \hat{i} + \hat{j} + \hat{k}$$

$$AN = \overrightarrow{ON} - \overrightarrow{OA} = \hat{i} + \hat{j} - \hat{k}$$

$\therefore \quad \overrightarrow{OP} \cdot \overrightarrow{AN} = 1 + 1 - 1 = 1$

$$|\overrightarrow{OP}| = |\overrightarrow{AN}| = \sqrt{1 + 1 + 1} = \sqrt{3}$$

If θ be the angle between $\overrightarrow{OP}$ and $\overrightarrow{AN}$, then

$$\cos\theta = \frac{\overrightarrow{OP}\cdot\overrightarrow{AN}}{\sqrt{3}\cdot\sqrt{3}} = \frac{1+1-1}{\sqrt{3}\cdot\sqrt{3}} = \frac{1}{3}$$

$$\therefore \quad \theta = \cos^{-1}\left(\frac{1}{3}\right)$$

VECTOR (OR CROSS) PRODUCT OF TWO VECTORS

The vector product of two vectors $\vec{a}$ and $\vec{b}$ is denoted by $\vec{a}\times\vec{b}$ (read as $\vec{a}$ cross $\vec{b}$) and defined as

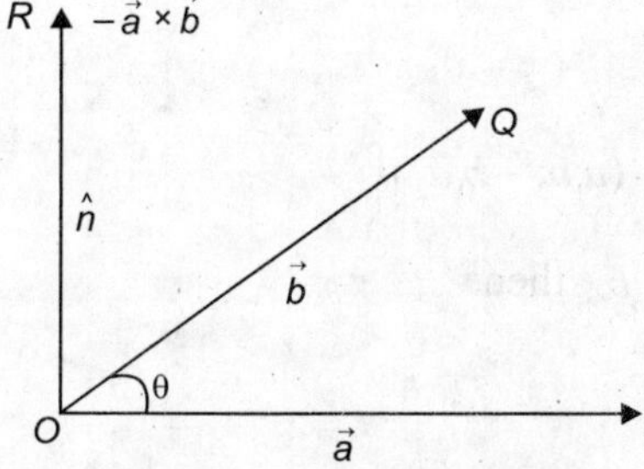

$$\vec{a}\times\vec{b} = |\vec{a}||\vec{b}|\sin\theta\,\hat{n}$$

where θ is the angle between $\vec{a}$ and $\vec{b}$ and $\hat{n}$ is a unit vector normal to the plane $\vec{a}$ and $\vec{b}$ such that the vectors $\vec{a}, \vec{b}$ and $\hat{n}$ form a right-handed system.

Properties

(1) Vector product is not commutative

If $\vec{a}$ and $\vec{b}$ are two vectors, then

$$\vec{a}\times\vec{b} \neq \vec{b}\times\vec{a}$$

But

$$\vec{a}\times\vec{b} = -\vec{b}\times\vec{a}$$

(2) Vector product is distributive with respect to vector addition.

If $\vec{a}, \vec{b}, \vec{c}$ are three vectors, then

$$\vec{a}\times(\vec{b}+\vec{c}) = \vec{a}\times\vec{b} + \vec{a}\times\vec{c}$$

(3) Vector product of two parallel vectors is zero

If two vectors $\vec{a}$ and $\vec{b}$ are parallel, then θ, the angle between them is '0' or π so that sin θ = 0

$$\therefore \qquad \vec{a}\times\vec{b} = |\vec{a}||\vec{b}|\sin 0 = 0$$

$$\therefore \qquad \vec{a}\times\vec{b} = 0$$

In particular, $\qquad \vec{a}\times\vec{a} = 0$

(4) If $\hat{i}$, $\hat{j}$, $\hat{k}$ are the unit vectors along the co-ordinate axes, then

$$\hat{i} \times \hat{i} = \hat{j} \times \hat{j} = \hat{k} \times \hat{k} = 0$$

and $$\hat{i} \times \hat{j} = \hat{k}, \ \hat{j} \times \hat{k} = \hat{i}, \ \hat{k} \times \hat{i} = \hat{j}$$

Also $$\hat{j} \times \hat{i} = -\hat{k}, \ \hat{k} \times \hat{j} = -\hat{i}, \ \hat{i} \times \hat{k} = -\hat{j}$$

(5) If $\vec{a} = a_1\hat{i} + a_2\hat{j} + a_3\hat{k}$ and $\vec{b} = b_1\hat{i} + b_2\hat{j} + b_3\hat{k}$

Then $$\vec{a} \times \vec{b} = \begin{vmatrix} \hat{i} & \hat{j} & \hat{k} \\ a_1 & a_2 & a_3 \\ b_1 & b_2 & b_3 \end{vmatrix}$$

$$= (a_2 b_3 - b_2 a_3)\hat{i} - (a_1 b_3 - b_1 a_3)\hat{j} + (a_1 b_2 - b_1 a_2)\hat{k}$$

(6) If θ is the angle between the vectors $\vec{a}$ and $\vec{b}$, then

$$\sin\theta = \frac{|\vec{a} \times \vec{b}|}{|\vec{a}||\vec{b}|}$$

(7) If $\vec{a}$ and $\vec{b}$ are two vectors, then a unit vector normal to these vectors is

$$\hat{n} = \frac{\vec{a} \times \vec{b}}{|\vec{a} \times \vec{b}|}$$

Examples

(1) Given $\vec{a} = 2\hat{i} + 2\hat{j} - \hat{k}$, and $\vec{b} = 6\hat{i} - 3\hat{j} + 2\hat{k}$ find $\vec{a} \times \vec{b}$ and a unit vector perpendicular to both $\vec{a}$ and $\vec{b}$. Also determine the sine of the angle between $\vec{a}$ and $\vec{b}$.

Solution: $\vec{a} = 2\hat{i} + 2\hat{j} - \hat{k}$ and $\vec{b} = 6\hat{i} - 3\hat{j} + 2\hat{k}$

$$\therefore \qquad \vec{a} \times \vec{b} = \begin{vmatrix} \hat{i} & \hat{j} & \hat{k} \\ 2 & 2 & -1 \\ 6 & -3 & 2 \end{vmatrix} = \hat{i} - 10\hat{j} - 18\hat{k}$$

$$\therefore \qquad |\vec{a}\times\vec{b}| = \sqrt{1+100+324} = \sqrt{425} = 5\sqrt{17}$$

$\therefore$ Unit vector $\hat{n}\perp$ to both $\vec{a}$ and $\vec{b}$

$$= \frac{\vec{a}\times\vec{b}}{|\vec{a}\times\vec{b}|} = \frac{1}{5\sqrt{17}}(\hat{i}-10\hat{j}-18\hat{k})$$

Also
$$|\vec{a}| = \sqrt{2^2+2^2+1^2} = 3$$

$$|\vec{b}| = \sqrt{6^2+3^2+2^2} = 7$$

If θ is the angle between $\vec{a}$ and $\vec{b}$, then,

$$\sin\theta = \frac{|\vec{a}\times\vec{b}|}{|\vec{a}||\vec{b}|} = \frac{5\sqrt{17}}{3\times 7} = \frac{5\sqrt{17}}{21}$$

$$\therefore \quad \theta = \sin^{-1}\left(\frac{5\sqrt{17}}{21}\right)$$

(2) Find a unit vector normal to both the vectors $4\hat{i}-\hat{j}+3\hat{k}$ and $-2\hat{i}+\hat{j}-2\hat{k}$.

Solution: Let $\vec{a} = 4\hat{i}-\hat{j}+3\hat{k}$ and $\vec{b} = -2\hat{i}+\hat{j}-2\hat{k}$

$$\therefore \quad \vec{a}\times\vec{b} = \begin{vmatrix} \hat{i} & \hat{j} & \hat{k} \\ 4 & -1 & 3 \\ -2 & 1 & -2 \end{vmatrix} = (+2-3)\hat{i}-(-8+6)\hat{j}+(4-2)\hat{k}$$

$$= -\hat{i}-2\hat{j}+2\hat{k}$$

$\therefore$ Unit vector normal to $\vec{a}$ and $\vec{b}$

$$= \frac{-\hat{i}-2\hat{j}+2\hat{k}}{\sqrt{1+4+4}} = \frac{1}{3}(-\hat{i}-2\hat{j}+2\hat{k})$$

(3) Find a unit vector normal to the vectors $\vec{a} = 3\hat{i}-2\hat{j}+4\hat{k}$ and $\vec{b} = \hat{i}+\hat{j}-2\hat{k}$. Also find the sine of the angle between them.

Solution: Let $\vec{a} = 3\hat{i}-2\hat{j}+4\hat{k}$ and $\vec{b} = \hat{i}+\hat{j}-2\hat{k}$.

$$\therefore \quad \vec{a}\times\vec{b} = \begin{vmatrix} \hat{i} & \hat{j} & \hat{k} \\ 3 & -2 & 4 \\ 1 & 1 & -2 \end{vmatrix} = (4-4)\hat{i} - (-6-4)\hat{j} + (3+2)\hat{k}$$

$$= 0\hat{i} + 10\hat{j} + 5\hat{k}$$

$\therefore$ Unit vector normal to the vectors $\vec{a}$ and $\vec{b}$

$$= \frac{0\hat{i}+10\hat{j}+5\hat{k}}{\sqrt{10^2+5^2}} = \frac{10\hat{j}+5\hat{k}}{5\sqrt{5}} = \frac{2}{\sqrt{5}}\hat{j} + \frac{1}{\sqrt{5}}\hat{k}$$

Now,
$$|\vec{a}| = \sqrt{3^2+2^2+4^2} = \sqrt{29}$$

$$|\vec{b}| = \sqrt{1^2+1^2+2^2} = \sqrt{6}$$

and
$$|\vec{a}\times\vec{b}| = \sqrt{10^2+5^2} = 5\sqrt{5}$$

$\therefore$
$$\sin\theta = \frac{|\vec{a}\times\vec{b}|}{|\vec{a}||\vec{b}|} = \frac{5\sqrt{5}}{\sqrt{29}\sqrt{6}}$$

$\therefore$
$$\theta = \sin^{-1}\left(\frac{5\sqrt{5}}{\sqrt{29}\sqrt{6}}\right)$$

(4) If $\vec{A} = \hat{i} - 2\hat{j} - 3\hat{k}$, $\vec{B} = 2\hat{i} + \hat{j} - \hat{k}$, $\vec{C} = \hat{i} + 3\hat{j} - \hat{k}$.

Find $(\vec{A}\times\vec{B})\times(\vec{B}\times\vec{C})$.

Solution: Let $\vec{A} = \hat{i} - 2\hat{j} - 3\hat{k}$, $\vec{B} = 2\hat{i} + \hat{j} - \hat{k}$, $\vec{C} = \hat{i} + 3\hat{j} - \hat{k}$.

$$\vec{A}\times\vec{B} = \begin{vmatrix} \hat{i} & \hat{j} & \hat{k} \\ 1 & -2 & -3 \\ 2 & 1 & -1 \end{vmatrix} = (2+3)\hat{i} - (1+6)\hat{j} + (1+4)\hat{k}$$

$$= 5\hat{i} - 7\hat{j} + 5\hat{k}$$

$$\vec{B}\times\vec{C} = \begin{vmatrix} \hat{i} & \hat{j} & \hat{k} \\ 2 & 1 & -1 \\ 1 & 3 & -1 \end{vmatrix} = (-1+3)\hat{i} - (-2+1)\hat{j} + (6-1)\hat{k}$$

$$= 2\hat{i} + \hat{j} + 5\hat{k}$$

$$(\vec{A} \times \vec{B}) \times (\overline{B} \times \overline{C}) = \begin{vmatrix} \hat{i} & \hat{j} & \hat{k} \\ 5 & -7 & 5 \\ 2 & 1 & 5 \end{vmatrix}$$

$$= (-35 - 5)\hat{i} - (25 - 10)\hat{j} + (5 + 14)\hat{k}.$$

$$= -40\hat{i} - 15\hat{j} + 19\hat{k}$$

SCALAR TRIPLE PRODUCT

Let $\vec{a}, \overline{b}, \vec{c}$ be any three vectors, then the scalar or dot product of $\vec{a} \times \vec{b}$ with $\vec{c}$ is called the scalar product of three vectors $\overline{a}, \overline{b}, \vec{c}$ and is denoted by $(\overline{a} \times \overline{b}) \cdot \vec{c}$ or $[\vec{a}\ \vec{b}\ \vec{c}]$.

Geometrically, the product $(\overline{a} \times \overline{b}) \cdot \vec{c}$ represents numerically the volume of the parallelopiped having $\vec{a}, \vec{b}, \vec{c}$ as coterminous edges.

Properties

(1) In a scalar triple product, the position of dot and cross can be interchanged, provided the cycle order of the vectors is maintained. However, if the cycle order is reversed, the sign of the scalar triple product changes.

i.e., $$[\overline{a}\ \overline{b}\ \overline{c}] = [\overline{b}\ \overline{c}\ \overline{a}] = [\overline{c}\ \overline{a}\ \overline{b}]$$

and $$[\overline{a}\ \overline{b}\ \overline{c}] = -[\overline{b}\ \overline{a}\ \overline{c}]$$

(2) The condition for three vectors to be coplanar is that their scalar triple product is zero.

If $\vec{a}, \overline{b}, \vec{c}$ are coplanar, then $\vec{a} \times \vec{b}$ is a vector perpendicular to the plane containing $\vec{a}, \overline{b}$ and $\vec{c}$.

$$\therefore \qquad \vec{a} \times \vec{b} \text{ is perpendicular to } \vec{c}$$

$$\therefore \qquad (\overline{a} \times \overline{b}) \cdot \overline{c} = 0$$

(3) If $\vec{a} = a_1\hat{i} + a_2\hat{j} + a_3\hat{k}, \overline{b} = b_1\hat{i} + b_2\hat{j} + b_3\hat{k}, \vec{c} = c_1\hat{i} + c_2\hat{j} + c_3\hat{k}$

then

$$\vec{a} \cdot (\overline{b} \times \overline{c}) = \vec{a} \cdot \begin{vmatrix} i & j & k \\ b_1 & b_2 & b_3 \\ c_1 & c_2 & c_3 \end{vmatrix}$$

$$= (a_1\hat{i} + a_2\hat{j} + a_3\hat{k}) \cdot [(b_2c_3 - b_3c_2)\hat{i} - (b_1c_3 - c_1b_3)\hat{j} + (b_1c_2 - c_1b_2)\hat{k}]$$

$$= a_1(b_2c_3 - b_3c_2) - a_2(b_1c_3 - c_1b_3) + a_3(b_1c_2 - c_1b_2)$$

$$= \begin{vmatrix} a_1 & a_2 & a_3 \\ b_1 & b_2 & b_3 \\ c_1 & c_2 & c_3 \end{vmatrix}$$

(4) If $\hat{i}, \hat{j}, \hat{k}$ are the unit vectors along the co-ordinate axes, then

$$[\hat{i} \ \hat{j} \ \hat{k}] = (\hat{i} \times \hat{j}) \times \hat{k} = \hat{k} \times \hat{k} = 1$$

Examples

(1) Prove that $[\overline{a} + \overline{b} \ \overline{b} + \overline{c} \ \overline{c} + \overline{a}] = 2[\overline{a} \ \overline{b} \ \overline{c}]$

Solution: $[\vec{a} + \overline{b}, \overline{b} + \overline{c}, \overline{c} + \overline{a}] = (\overline{a} + \overline{b}) \cdot [(b \times \overline{c}) \times (\overline{c} \times \overline{a})]$

$$= (\vec{a} + \vec{b}) \cdot [\overline{b} \times \overline{c} + \overline{b} \times \vec{a} + \overline{c} \times \vec{c} + \overline{c} \times \overline{a}]$$

$$= (\vec{a} + \vec{b}) \cdot [\overline{b} \times \overline{c} + \overline{b} \times \vec{a} + \overline{c} \times \overline{a}] \qquad \because \overline{c} \times \overline{c} = 0$$

$$= \overline{a} \cdot (\overline{b} \times \overline{c}) + (\vec{a} \cdot \overline{b} \times \vec{a}) + (\vec{a} \cdot \overline{c} \times \vec{a}) + (\overline{b} \cdot \overline{b} \times \overline{c})$$

$$+ (\overline{b} \cdot \overline{b} \times \overline{a}) + (\overline{b} \cdot \overline{c} \times \vec{a})$$

$$= [\vec{a} \ \overline{b} \ \overline{c}] + 0 + 0 + 0 + 0 + [\overline{b} \ \overline{c} \ \vec{a}]$$

$$= [\overline{a} \ \overline{b} \ \overline{c}] + [\overline{a} \ \overline{b} \ \overline{c}] \qquad (\because \text{ scalar product is distributive})$$

$$= 2[\overline{a} \ \overline{b} \ \overline{c}]$$

(2) If $\vec{a}, \overline{b}, \vec{c}$ are the position vectors of the points A, B, C, prove that the vector

$\vec{a} \times \overline{b} + \overline{b} \times \overline{c} + \overline{c} \times \vec{a}$ is perpendicular to the plane of triangle ABC.

Solution: $\overrightarrow{AB} = \overrightarrow{OB} - \overrightarrow{OA} = \overline{b} - \overline{a}$

and $(\vec{a} \times \vec{b} + \vec{b} \times c + \vec{c} \times \vec{a}) \cdot (\vec{b} - \vec{a})$

$$= [a\ \vec{b}\ \vec{b}] + [\vec{b}\ \vec{c}\ \vec{b}] + [\vec{c}\ \vec{a}\ \vec{b}] - [\vec{a}\ \vec{b}\ \vec{a}] - [\vec{b}\ c\ \vec{a}] - [\vec{c}\ \vec{a}\ \vec{a}]$$

$$= 0 + 0 + [\vec{c}\ \vec{a}\ \vec{b}] - 0 - [\vec{c}\ \vec{a}\ \vec{b}] - 0 = 0$$

$\therefore$ The given vector is perpendicular to $\overrightarrow{AB}$.

Similarly, we can prove that $\overrightarrow{BC}$ and $\overrightarrow{CA}$ are perpendicular to the given vector.

$\therefore$ The given vector is perpendicular to the given plane of triangle ABC.

(3) Prove that the vectors $\hat{i} - 2\hat{j} + 3\hat{k},\ -2\hat{i} + 3\hat{j} - 4\hat{k}$ and $\hat{i} - 3\hat{j} + 5\hat{k}$ are coplanar.

Proof: Let $\vec{a} = \hat{i} - 2\hat{j} + 3\hat{k},\ \vec{b} = -2\hat{i} + 3\hat{j} - 4\hat{k}$ and $\vec{c} = \hat{i} - 3\hat{j} + 5\hat{k}$.

The vectors $\vec{a}, \vec{b}$ and $\vec{c}$ are co-planar if $\vec{a} \cdot (\vec{b} \times \vec{c}) = 0$.

Consider

$$\vec{a} \cdot (\vec{b} \times \vec{c}) = \begin{vmatrix} 1 & -2 & 3 \\ -2 & 3 & -4 \\ 1 & -3 & 5 \end{vmatrix} = 1(15 - 12) + 2(-10 + 4) + 3(6 - 3)$$

$$= 3 - 12 + 9 = 0$$

Since the scalar triple product is zero, the vectors $\vec{a}, \vec{b}, \vec{c}$ are coplanar.

(4) Find the constant a so that the vectors $2\hat{i} - \hat{j} + \hat{k}$, $\hat{i} + 2\hat{j} - 3\hat{k}$ and $3\hat{i} + a\hat{j} + 5\hat{k}$ are coplanar.

Solution: Let $\vec{a} = 2\hat{i} - \hat{j} + \hat{k},\ \vec{b} = \hat{i} + 2\hat{j} - 3\hat{k}$ and $\vec{c} = 3\hat{i} + a\hat{j} + 5\hat{k}$ be three vectors.

These vectors are coplanar if $\vec{a} \cdot (\vec{b} \times \vec{c}) = 0$

i.e.,
$$\begin{vmatrix} 2 & -1 & 1 \\ 1 & 2 & -3 \\ 3 & a & 5 \end{vmatrix} = 0$$

$$2(10 + 3a) + 1(5 + 9) + 1(a - 6) = 0$$

$\therefore$
$$20 + 6a + 14 + a - 6 = 0$$

$\therefore$
$$7a = -28 \qquad \therefore \boxed{a = -4}$$

(5) Show that the vectors $\vec{A} = \hat{i} - 2\hat{j} + 3\hat{k}$, $\vec{B} = 2\hat{i} + \hat{j} + \hat{k}$ and $\vec{C} = 3\hat{i} + 4\hat{j} - \hat{k}$ are coplanar.

Solution: The vectors $\vec{A}, \vec{B}, \vec{C}$ are coplanar if

$$\vec{A} \cdot (\vec{B} \times \vec{C}) = 0$$

Consider

$$\vec{A} \cdot (\vec{B} \times \vec{C}) = \begin{vmatrix} 1 & -2 & 3 \\ 2 & 1 & 1 \\ 3 & 4 & -1 \end{vmatrix}$$

$$= 1(-1-4) + 2(-2-3) + 3(8-3)$$

$$= -5 - 10 + 15$$

$$= 0$$

Since scalar triple product is zero, the vectors $\vec{A}, \vec{B}$ and $\vec{C}$ are coplanar.

VECTOR TRIPLE PRODUCT

If $\vec{a}, \vec{b}, \vec{c}$ are any three vectors, then the cross product of $\vec{a} \times \vec{b}$ with $\vec{c}$ is called the vector triple product of the three vectors and is defined as

$$(\vec{a} \times \vec{b}) \times \vec{c} = (\vec{a} \cdot \vec{c})\vec{b} - (\vec{b} \cdot \vec{c})\vec{a}$$

or $$\vec{a} \times (\vec{b} \times \vec{c}) = (\vec{a} \cdot \vec{c})\vec{b} - (\vec{a} \cdot \vec{b})\vec{c}$$

Theorem: Prove that

$$\vec{a} \times (\vec{b} \times \vec{c}) = (\vec{a} \cdot \vec{c})\vec{b} - (\vec{a} \cdot \vec{b})\vec{c}$$

Proof: Without loss of generality choose the vectors as $\vec{a} = a_1\hat{i}$, $\vec{b} = b_1\hat{i} + b_2\hat{j}$ and $\vec{c} = c_1\hat{i} + c_2\hat{j} + c_3\hat{k}$.

Now, $$\vec{b} \times \vec{c} = \begin{vmatrix} \hat{i} & \hat{j} & \hat{k} \\ b_1 & b_2 & 0 \\ c_1 & c_2 & c_3 \end{vmatrix} = b_2 c_3 \hat{i} - b_1 c_3 \hat{j} + (b_1 c_2 - c_1 b_2)\hat{k}$$

$$\therefore \qquad \vec{a} \times (\vec{b} \times \vec{c}) = \begin{vmatrix} i & j & k \\ a_1 & 0 & 0 \\ b_2 c_3 & -b_1 c_3 & b_1 c_2 - c_1 b_2 \end{vmatrix}$$

$$= \hat{i}(0 - 0) - \hat{j} a_1 (b_1 c_2 - c_1 b_2) + \hat{k}(a_1)(-b_1 c_3)$$

$$= a_1 c_1 b_2 \hat{j} - a_1 b_1 c_2 \hat{j} - a_1 b_1 c_3 \hat{k} \qquad (1)$$

Also $\qquad \vec{a} \cdot \vec{c} = a_1 \hat{i} \cdot (c_1 \hat{i} + c_2 \hat{j} + c_3 \hat{k}) = a_1 c_1$

$$\therefore \qquad (\vec{a} \cdot \vec{c})\vec{b} = a_1 c_1 b_1 \hat{i} + a_1 c_1 b_2 \hat{j}$$

and $\qquad \vec{a} \cdot \vec{b} = a_1 \hat{i} \cdot (b_1 \hat{i} + b_2 \hat{j}) = a_1 b_1$

$$\therefore \qquad (\vec{a} \cdot \vec{b})\vec{c} = a_1 b_1 c_1 \hat{i} + a_1 b_1 c_2 \hat{j} + a_1 b_1 c_3 \hat{k}$$

$$\therefore \qquad (\vec{a} \cdot \vec{c})\vec{b} - (\vec{a} \cdot \vec{b})\vec{c} = a_1 c_1 b_2 \hat{j} - a_1 b_1 c_2 \hat{j} - a_1 b_1 c_3 \hat{k} \qquad (2)$$

From (1) and (2)

$$\vec{a} \times (\vec{b} \times \vec{c}) = (\vec{a} \cdot \vec{c})\vec{b} - (\vec{a} \cdot \vec{b})\vec{c}$$

Examples

(1) Prove that $\vec{a} \times (\vec{b} \times \vec{c}) + b \times (\vec{c} \times \vec{a}) + \vec{c} \times (\vec{a} \times \vec{b}) = 0$.

Proof: Consider $\quad \vec{a} \times (\vec{b} \times \vec{c}) + \vec{b} \times (\vec{c} \times \vec{a}) + \vec{c} \times (\vec{a} \times \vec{b})$

$$= (\vec{a} \cdot \vec{c})\vec{b} - (\vec{a} \cdot \vec{b})\vec{c} + (\vec{b} \cdot a)\vec{c} - (\vec{b} \cdot c)\vec{a} + (\vec{c} \cdot \vec{b})\vec{a} - (\vec{c} \cdot \vec{a})\vec{b}$$

$$= 0 \qquad\qquad (\because \vec{a} \cdot \vec{c} = \vec{c} \cdot \vec{a}, \text{ etc.})$$

(2) Prove that

$$[\vec{b} \times \vec{c}, \vec{c} \times \vec{a}, \vec{a} \times \vec{b}] = [\vec{a}\ \vec{b}\ \vec{c}]^2$$

Proof: Consider

$$[\vec{b} \times \vec{c}, \vec{c} \times \vec{a}, \vec{a} \times \vec{b}] = (\vec{b} \times \vec{c}) \cdot (\vec{c} \times \vec{a}) \times (\vec{a} \times \vec{b})$$

$$= (\vec{b} \times \vec{c}) \cdot \{[\vec{c}\ \vec{a}\ \vec{b}]\vec{a} - [\vec{c}\ \vec{a}\ \vec{a}]\vec{b}\}$$

$$= (\vec{b} \times \vec{c}) \cdot \{[\vec{a}\ \vec{b}\ \vec{c}]\vec{a} - 0\}$$

$$= \overline{b} \times \overline{c} \cdot \overline{a} [\overline{a} \ \overline{b} \ \overline{c}]$$

$$= [\overline{a} \ \overline{b} \ \overline{c}][\overline{a} \ \overline{b} \ \overline{c}] = [\overline{a} \ \overline{b} \ \overline{c}]^2 .$$

(3) Prove that $\vec{a} \times \{\overline{b} \times (\overline{c} \times \overline{d})\} = \overline{b} \cdot \overline{d}(\overline{a} \times c) - (\overline{b} \cdot \overline{c})(\overline{a} \times \overline{d})$

Proof: Consider

$$\vec{a} \times \{\overline{b} \times (\overline{c} \times \overline{d})\} = \vec{a} \times \{(\overline{b} \cdot \overline{d})\vec{c} - (\overline{b} \cdot \overline{c})\vec{d}$$

$$= (\overline{b} \cdot \overline{d})(\vec{a} \times \overline{c}) - (\overline{b} \cdot \overline{c})(\vec{a} \times \overline{d}).$$

(4) If $\vec{a} = \hat{i} - 2\hat{j} - 3\hat{k}, \vec{b} = 2\hat{i} + \hat{j} - \hat{k}, \vec{c} = \hat{i} + 3\hat{j} - \hat{k},$ find

(i) $$\vec{a} \times (\overline{b} \times \overline{c})$$

(ii) $$(\vec{a} \times \overline{b}) \times (\overline{b} \times \overline{c})$$

Solution: $\vec{a} = \hat{i} - 2\hat{j} - 3\hat{k}, \vec{b} = 2\hat{i} + \hat{j} - \hat{k}, \vec{c} = \hat{i} + 3\hat{j} - \hat{k}$

(i) $$\overline{b} \times \overline{c} = \begin{vmatrix} \hat{i} & \hat{j} & \hat{k} \\ 2 & 1 & -1 \\ 1 & 3 & -1 \end{vmatrix} = 2\hat{i} + \hat{j} + 5\hat{k}$$

$\therefore$ $$\vec{a} \times (\overline{b} \times \overline{c}) = \begin{vmatrix} \hat{i} & \hat{j} & \hat{k} \\ 1 & -2 & -3 \\ 2 & 1 & 5 \end{vmatrix} = 7\hat{i} - 11\hat{j} + 5\hat{k}$$

(ii) $$\vec{a} \times \overline{b} = \begin{vmatrix} \hat{i} & \hat{j} & \hat{k} \\ 1 & -2 & -3 \\ 2 & 1 & -1 \end{vmatrix} = 5\hat{i} - 5\hat{j} + 5\hat{k}$$

$\therefore$ $$(\vec{a} \times \overline{b}) \times (\overline{b} \times \overline{c}) = \begin{vmatrix} \hat{i} & \hat{j} & \hat{k} \\ 5 & -5 & 5 \\ 2 & 1 & 5 \end{vmatrix}$$

$$= (-25 - 5)\hat{i} - (25 - 10)\hat{j} + (5 + 10)\hat{k}$$

$$= -30\hat{i} - 15\hat{j} + 15\hat{k}$$

(5) Constant forces $\vec{P} = 2\hat{i} - 5\hat{j} + 6\hat{k}$ and $\vec{Q} = -\hat{i} + 2\hat{j} - \hat{k}$ act on a particle. Determine the work done when the particle is displaced from A to B, the position vectors of A and B being $4\hat{i} - 3\hat{j} - 2\hat{k}$ and $6\hat{i} + \hat{j} - 3\hat{k}$ respectively.

Solution: Resultant force

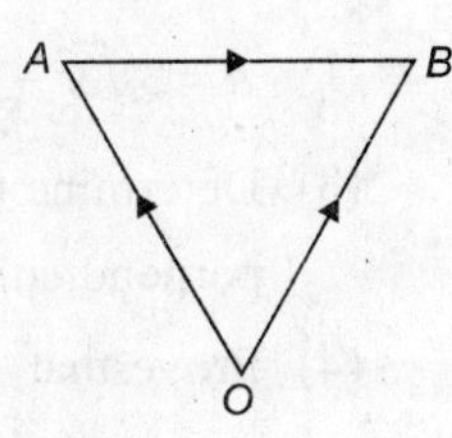

$$\vec{F} = \vec{P} + \vec{Q} = \hat{i} - 3\hat{j} + 5\hat{k}$$

and

$$\overrightarrow{AB} = \overrightarrow{OB} - \overrightarrow{OA}$$

$$= (6\hat{i} + \hat{j} - 3\hat{k}) - (4\hat{i} - 3\hat{j} - 2\hat{k})$$

$$= 2\hat{i} + 4\hat{j} - \hat{k}$$

$$\text{Work done} = \vec{F}.\overrightarrow{AB}$$

$$= (\hat{i} - 3\hat{j} + 5\hat{k}).(2\hat{i} + 4\hat{j} - \hat{k})$$

$$= 2 - 12 - 5 = -15 \text{ units}$$

(6) Find the torque about the point $2\hat{i} + \hat{j} - \hat{k}$ of a force represented by $4\hat{i} + \hat{k}$ acting through the point $\hat{i} - \hat{j} + 2\hat{k}$.

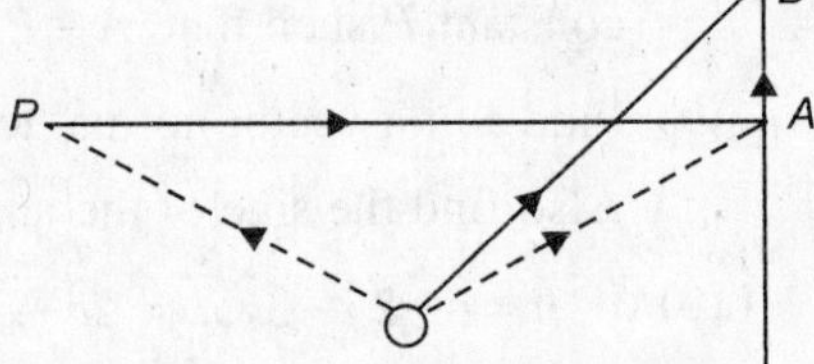

Let O be the origin and P be the point moment about which the force $\overrightarrow{AB}$ through A, is required.

$$\therefore \qquad \overrightarrow{OP} = 2\hat{i} + \hat{j} - \hat{k}$$

$$\overrightarrow{OA} = \hat{i} - \hat{j} + 2\hat{k} \text{ and } \overrightarrow{AB} = 4\hat{i} + \hat{k}.$$

Then $\overrightarrow{PA} = \overrightarrow{OA} - \overrightarrow{OP} = -\hat{i} - 2\hat{j} + 3\hat{k}$

$\therefore$ Moment of the force $\overrightarrow{AB}$ about P

$$= \overrightarrow{PA} \times \overrightarrow{AB} = (-\hat{i} - 2\hat{j} + 3\hat{k}) \times (4\hat{i} + \hat{k})$$

$$= \begin{vmatrix} \hat{i} & \hat{j} & \hat{k} \\ -1 & -2 & 3 \\ 4 & 0 & 1 \end{vmatrix}$$

$$= -2\hat{i} + 13\hat{j} + 8\hat{k}$$

$\therefore$ Magnitude of the moment $= \sqrt{4 + 169 + 64} = 15.4$

QUESTION BANK 3

(1) Define scalar product and cross product of two vectors. If
$\vec{a} = 2\hat{i} + 3\hat{j} - 5\hat{k}$ and $\vec{b} = 3\hat{i} - 5\hat{j} - 7\hat{k}$. Find $\vec{a} \cdot \vec{b}$ and $\vec{a} \times \vec{b}$.

(2) Show that the position vectors of the vertices of a triangle
$\vec{a} = 3\sqrt{3}\hat{i} - 3\hat{j}, \vec{b} = 6\hat{j}, \vec{c} = 3\sqrt{3}\hat{i} + 3\hat{j}$ form an isosceles triangle.

(3) Determine the value of t so that $\vec{a} = 2\hat{i} + t\hat{j} + \hat{k}$ and $\vec{b} = 4\hat{i} - 2\hat{j} - 2\hat{k}$ are perpendicular.

(4) Prove that $[\vec{A} + \vec{B}, \vec{B} + \vec{C}, \vec{C} + \vec{A}] = 2[\vec{A}\ \vec{B}\ \vec{C}]$.

(5) Find the constant a so that the vectors $2\hat{i} - \hat{j} + \hat{k}, \hat{i} + 2\hat{j} - 3\hat{k}$ and $3\hat{i} + a\hat{j} + 5\hat{k}$ are coplanars.

(6) If $\vec{a}, \vec{b}, \vec{c}$ are any three vectors prove that
$$[\vec{b} \times \vec{c}, \vec{c} \times \vec{a}, \vec{a} \times \vec{b}] = [\vec{a}\ \vec{b}\ \vec{c}]^2$$

(7) Prove that $\vec{a} \times (\vec{b} \times \vec{c}) = (\vec{a} \cdot \vec{c})\vec{b} - (\vec{a} \cdot \vec{b})\vec{c}$.

(8) If $\vec{A} = \hat{i} + 2\hat{j} + 3\hat{k}, \vec{B} = -\hat{i} + 2\hat{j} + \hat{k}$ and $\vec{c} = 3\hat{i} + \hat{j}$, find the value of the constant P such that $\vec{A} + P\vec{B}$ is perpendicular to $\vec{C}$.

(9) Find a unit vector normal to the plane of $\vec{a} = 3\hat{i} - 2\hat{j} + 4\hat{k}, \vec{b} = \hat{i} + \hat{j} - 2\hat{k}$. Also find the sine of the angle between them.

(10) If $\vec{a} = \hat{i} - 2\hat{j} - 3\hat{k}, \vec{b} = 2\hat{i} + \hat{j} - \hat{k}, \vec{c} = \hat{i} + 3\hat{j} - \hat{k}.$, then find
$(\vec{a} \times \vec{b}) \times (\vec{b} \times \vec{c})$.

(11) Given $\vec{a} = 2\hat{i} + 2\hat{j} - \hat{k}, \vec{b} = 6\hat{i} + 3\hat{j} + 2\hat{k}$, find $\vec{a} \times \vec{b}$ and the unit vector perpendicular to both $\vec{a}$ and $\vec{b}$. Also determine the sine of the angle between $\vec{a}$ and $\vec{b}$.

(12) Determine the value of λ so that $\vec{a} = 2\hat{i} + \lambda\hat{j} + \hat{k}$ and $\vec{b} = 4\hat{i} - 2\hat{j} - 2\hat{k}$ are perpendicular.

(13) Show that the vectors $\vec{a} = 3\hat{i} - 2\hat{j} + \hat{k}, \vec{b} = \hat{i} - 3\hat{j} + 5\hat{k}$ and
$\vec{c} = 2\hat{i} + \hat{j} - 4\hat{k}$ form a right handed triangle.

(14) Find the projection of the vector $\hat{i} - 2\hat{j} + \hat{k}$ on $4\hat{i} - 4\hat{j} + 7\hat{k}$.

(15) Constant force $\vec{P} = 2\hat{i} - 5\hat{j} + 6\hat{k}$ and $\vec{Q} = -\hat{i} + 2j - \hat{k}$ act on a particle. Determine the work done when the particle is displaced from A to B, the

position vectors of A and B being $4\hat{i} - 3\hat{j} - 2\hat{k}$ and $6\hat{i} + \hat{j} - 3\hat{k}$ respectively.

Ans: Work done $= \vec{F} \cdot \overrightarrow{AB} = -15$ units

(16) Evaluate $\vec{a} \cdot (\vec{b} \times \vec{c})$ if $\vec{a} = 2\hat{i} - 3\hat{j}$, $\vec{b} = \hat{i} + \hat{j} - \hat{k}$ and $\vec{c} = 3\hat{i} - \hat{k}$.

(17) Prove that the vectors $\hat{i} - 2\hat{j} + 3\hat{k}$, $-2\hat{i} + 3\hat{j} - 4\hat{k}$ and $\hat{i} - 3\hat{j} + 5\hat{k}$ are coplanar.

(18) Prove that $(\vec{a} \times \vec{b}) \cdot (\vec{b} \times \vec{c}) \times (\vec{c} \times a) = (\vec{a} \cdot \vec{b} \times \vec{c})$

(19) Show that $(\vec{a} \times \vec{b}) \times (\vec{c} \times \vec{d}) + (\vec{a} \times \vec{c}) \times (\vec{d} \times \vec{b}) + (\vec{a} \times \vec{d}) \times (\vec{b} \times \vec{c})$ is parallel to the vector $\vec{a}$.

(20) Prove that $(\vec{a} + \vec{b} + \vec{c}) \times (\vec{b} + \vec{c}) \cdot \vec{c} = \vec{a} \cdot (\vec{b} \times \vec{c})$.

UNIT 4

Vector Calculus

SCALAR AND VECTOR POINT FUNCTION

If to each point t of the region E there corresponds a f, (say) then f is called a scalar point function in E. The region E is called a scalar field. For example, the temperature at any instant is a scalar point function of the variable time.

If to each point of the region E there corresponds a definite vector say $\vec{F}$ then $\overline{F}$ is called a vector point function in E. The region E is called the vector field.

DIFFERENTIATION OF VECTORS

Consider a vector $\vec{R}$ which is the function of the variable t, i.e., $\vec{R} = \vec{F}(t)$ then

$$\vec{R} + \delta\vec{R} = \vec{F}(t + \delta t)$$

$$\therefore \qquad \delta\vec{R} = \vec{F}(t + \delta t) - \vec{F}(t)$$

Divide by dt,

$$\therefore \qquad \frac{\delta\vec{R}}{\delta t} = \frac{\vec{F}(t + \delta t) - \vec{F}(t)}{\delta t}$$

$$\therefore \qquad \lim_{\delta t \to 0} \frac{\delta\vec{R}}{\delta t} = \lim_{\delta t \to 0} \frac{\vec{F}(t + \delta t) - \vec{F}(t)}{\delta t}$$

The limit of this, if it exists, is called the *derivative of* $\vec{F}$ and is denoted by $F'(t)$ or $\dfrac{d\vec{R}}{dt}$.

If the scalar variable t represents the time and $\delta\vec{R}$ represents the small displacement in the instant δt, then $\dfrac{d\vec{R}}{dt} = \lim\limits_{\delta t \to 0} \dfrac{\delta \vec{R}}{\delta t}$ defines the *velocity* $\vec{V}$ and $\dfrac{d\vec{v}}{dt} = \dfrac{d^2 \vec{R}}{dt^2}$ defines

acceleration $\overline{A}$.

Theorem: Let $\overline{a}, \overline{b}$, and $\overline{c}$ be three vector point functions of the scalar variable t then

(1) $\dfrac{d}{dt}(\overline{a} \pm \overline{b}) = \dfrac{d\overline{a}}{dt} \pm \dfrac{d\overline{b}}{dt}$

(2) $\dfrac{d}{dt}(\vec{a} \pm \overline{b}) = \vec{a} \cdot \dfrac{d\overline{b}}{dt} + \dfrac{d\vec{a}}{dt} \times \overline{b}$

(3) $\dfrac{d}{dt}(\vec{a} \pm \overline{b}) = \vec{a} \times \dfrac{d\overline{b}}{dt} + \dfrac{d\vec{a}}{dt} \times \vec{b}$

(4) $\dfrac{d}{dt}(\phi \vec{a}) = \phi \dfrac{d\vec{a}}{dt} + \dfrac{d\phi}{dt} \vec{a}$ where ϕ is a scalar point function.

(5) $\dfrac{d}{dt}[\overline{a}\ \overline{b}\ \overline{c}] = \left[\dfrac{d\overline{a}}{dt}\ \overline{b}\ \overline{c}\right] + \left[\overline{a}\ \dfrac{d\overline{b}}{dt}\ \overline{c}\right] + \left[\overline{a}\ \overline{b}\ \dfrac{d\overline{c}}{dt}\right]$

(6) $\dfrac{d}{dt}\{\vec{a} \times (\overline{b} \times \overline{c})\} = \dfrac{d\vec{a}}{dt} \times (\overline{b} \times \overline{c}) + \vec{a} \times \left(\dfrac{db}{dt} \times \overline{c}\right) + \vec{a} \times \left(\overline{b} \times \dfrac{d\overline{c}}{dt}\right)$

Proofs:

(1) $\dfrac{d}{dt}(\overline{a} + \overline{b}) = \lim\limits_{\delta t \to 0} \dfrac{[(\overline{a} + \delta\overline{a}) + (b + \delta\overline{b}) - (\overline{a} + \overline{b})]}{\delta t}$

$\qquad\qquad = \lim\limits_{\delta t \to 0} \left(\dfrac{\delta\overline{a} + \delta\overline{b}}{\delta t}\right)$

$\qquad\qquad = \lim\limits_{\delta t \to 0} \dfrac{\delta\overline{a}}{\delta t} + \lim\limits_{\delta t \to 0} \dfrac{\delta\overline{b}}{\delta t}$

$\qquad\qquad = \dfrac{d\overline{a}}{dt} + \dfrac{d\overline{b}}{dt}$

Similarly, $\qquad\qquad \dfrac{d}{dt}(\overline{a} - \overline{b}) = \dfrac{d\overline{a}}{dt} - \dfrac{d\overline{b}}{dt}.$

(2) $\dfrac{d}{dt}(\vec{a}\cdot\vec{b}) = \lim\limits_{\delta t\to 0}\dfrac{[(\vec{a}+\delta\vec{a})\cdot(\vec{b}+\delta\vec{b})-\vec{a}\cdot\vec{b}]}{\delta t}$

$$= \lim_{\delta t\to 0}\frac{[(\vec{a}.\delta\vec{b}+\delta\vec{a}\cdot\vec{b}+\delta\vec{a}\cdot\delta\vec{b}]}{\delta t} \qquad (\because \vec{a}\cdot\vec{b}-\vec{a}\cdot\vec{b}=0)$$

$$= \lim_{\delta t\to 0}\left[a\cdot\frac{\delta\vec{b}}{\delta t}+\frac{\delta\vec{a}}{\delta t}\cdot b+\frac{\delta\vec{a}}{\delta t}\cdot\delta\vec{b}\right]$$

$$= \vec{a}\cdot\frac{d\vec{b}}{dt}+\frac{d\vec{a}}{dt}\cdot\vec{b} \quad (\because \delta b\to 0 \text{ as } \delta t\to 0)$$

$$= \vec{a}\cdot\frac{d\vec{b}}{dt}+\frac{d\vec{a}}{dt}\cdot\vec{b}$$

(3) $\dfrac{d}{dt}(\vec{a}\times\vec{b}) = \lim\limits_{\delta t\to 0}\dfrac{(\vec{a}+\delta\vec{a})\times(\vec{b}+\delta\vec{b})-\vec{a}\times\vec{b}}{\delta t}$

$$= \lim_{\delta t\to 0}\frac{\vec{a}\times\vec{b}+a\times\delta\vec{b}+\delta\vec{a}\times\vec{b}+\delta\vec{a}\times\delta\vec{b}-\vec{a}\times\vec{b}}{\delta t}$$

$$= \lim_{\delta t\to 0}\left[\vec{a}\times\frac{\delta\vec{b}}{\delta t}+\frac{\delta\vec{a}}{\delta t}\times\vec{b}+\frac{\delta\vec{a}}{\delta t}\times\delta\vec{b}\right]$$

$$= \vec{a}\times\frac{d\vec{b}}{dt}+\frac{d\vec{a}}{dt}\times\vec{b} \quad (\text{since } \delta t\to 0,\ \delta b\to 0 \text{ also})$$

$$\left(\because \frac{d\vec{a}}{dt}\times 0 = 0\right)$$

(4) $\dfrac{d}{dt}(\phi\vec{a}) = \lim\limits_{\delta t\to 0}\dfrac{(\phi+\delta\phi)(\vec{a}+\delta\vec{a})-\phi\vec{a}}{\delta t}$

$$= \lim_{\delta t\to 0}\frac{\phi\vec{a}+\phi\delta\vec{a}+\delta\phi\vec{a}+\delta\phi\delta\vec{a}-\phi\vec{a}}{\delta t}$$

$$= \lim_{\delta t\to 0}\left(\phi\frac{\delta\vec{a}}{\delta t}+\frac{\delta\phi}{\delta t}\vec{a}\right)+\lim_{\delta t\to 0}\frac{\delta\phi}{\delta t}\delta\vec{a}$$

$$= \phi\frac{d\vec{a}}{dt}+\frac{d\phi}{dt}\vec{a}+0 \quad (\because \delta\vec{a}\leftrightarrow 0 \text{ as } \delta t\to 0)$$

$$= a\frac{d\phi}{dt} + \frac{d\phi}{dt}\vec{a}$$

Similarly, we can prove,

(5) $\dfrac{d}{dt}[\bar{a}\ \bar{b}\ \bar{c}] = \left[\bar{a}\ \bar{b}\ \dfrac{d\bar{c}}{dt}\right] + \left[a\ \dfrac{d\bar{b}}{dt}\ c\right] + \left[\bar{a}\ \bar{b}\ \dfrac{d\bar{c}}{dt}\right]$

and

(6) $\dfrac{d}{dt}[\bar{a}\times\bar{b}\times\bar{c}] = \dfrac{d\bar{a}}{dt}\times(\bar{b}\times\bar{c}) + \bar{a}\times\left(\dfrac{d\bar{b}}{dt}\times\bar{c}\right) + \vec{a}\times\left(\bar{b}\times\dfrac{d\bar{c}}{dt}\right)$

Note that in scalar triple product and vector triple products the cyclic order of factors must be maintained.

(7) The derivative of a constant vector is a null vector.

(8) Derivative of a vector function in terms of its components:

Let $\vec{R}$ be a vector function of the scalar variable t and let, $\vec{R} = x\hat{i} + y\hat{j} + z\hat{k}$, then

$$\frac{d\vec{R}}{dt} = \frac{dx}{dt}\hat{i} + \frac{dy}{dt}\hat{j} + \frac{dz}{dt}\hat{k}$$

where $x,\ y,\ z$ are functions of t.

(9) If $\vec{F}(t)$ has a constant magnitude, then

$$\vec{F}\cdot\frac{d\vec{F}}{dt} = 0$$

(10) If $\vec{F}(t)$ has a constant direction, then

$$\vec{F}\cdot\frac{d\bar{F}}{dt} = 0$$

Examples

(1) If $\vec{a} = 5t^2\hat{i} + t\hat{j} - t^3\hat{k},\ \bar{b} = \sin t - \hat{i} - \cos t\hat{j}$

Find

(i) $\dfrac{d}{dt}(\vec{a}\cdot\bar{b})$

(ii) $\dfrac{d}{dt}(\bar{a}\times\bar{b})$

76   *Advanced Mathematics-II*

Solution: $\vec{a} = 5t^2\hat{i} + t\hat{j} - t^3\hat{k}, \ \vec{b} = \sin t - \hat{i} - \cos t\hat{j}$

(i) $\dfrac{d}{dt}(\vec{a} \cdot \vec{b}) = \vec{a}\,\dfrac{d\vec{b}}{dt} + \dfrac{d\vec{a}}{dt} \cdot \vec{b}$

$$= (5t^2\hat{i} + t\hat{j} - t^3\hat{k}) \cdot (\cos t\hat{i} + \sin t\hat{j})$$

$$+ (10t\hat{i} + \hat{j} - 3t^2\hat{k}) \cdot (\sin t\hat{i} - \cos t\hat{j})$$

$$= 5t^2\cos t + t\sin t + 10t\sin t - \cos t$$

$$= (5t^2 - 1)\cos t + (t + 10t)\sin t \ .$$

(ii) $\dfrac{d}{dt}(\vec{a} \times \vec{b}) = \vec{a} \times \dfrac{d\vec{b}}{dt} + \dfrac{d\vec{a}}{dt} \times \vec{b}$

$$= (5t^2 + t\hat{j} - t^3\hat{k}) \times (\cos t\hat{i} + \sin t\hat{j})$$

$$+ (10t\hat{i} + \hat{j} - 3t^2\hat{k}) \times (\sin t\hat{i} - \cos t\hat{j})$$

$$= \begin{vmatrix} \hat{i} & \hat{j} & \hat{k} \\ 5t^2 & t & -t^3 \\ \cos t & \sin t & 0 \end{vmatrix} + \begin{vmatrix} \hat{i} & \hat{j} & \hat{k} \\ 10t & 1 & -3t^2 \\ \sin t & -\cos t & 0 \end{vmatrix}$$

$$= i(0 + t^3\sin t) - \hat{j}(0 - t^3\cos t) + \hat{k}(5t^2\sin t - t\cos t)$$

$$+ i(0 + 3t^2\cos t) - \hat{j}(0 - 3t^2\sin t) + \hat{k}(-10t\cos t - \sin t)$$

$$= (t^3\sin t + 3t^2\cos t)\hat{i} + (t^3\cos t + 3t^2\sin t)\hat{j}$$

$$+ \{(5t^2 - 1)\sin t - (10t + 1)\cos t\}\hat{k} \ .$$

(2) If $\dfrac{d\vec{u}}{dt} = \vec{w} \times \vec{u}$ and $\dfrac{d\vec{v}}{dt} = \vec{w} \times \vec{v}$, prove that $\dfrac{d}{dt}(\vec{u} \times \vec{v}) = \vec{w} \times (\vec{u} \times \vec{v})$.

Proof: $\dfrac{d}{dt}(\vec{u} \times \vec{v}) = \vec{u} \times \dfrac{d\vec{v}}{dt} + \dfrac{d\vec{u}}{dt} \times \vec{v}$

$$= \vec{u} \times (\vec{w} \times \vec{v}) + (\vec{w} \times \vec{u}) \times v$$

$$= (\vec{u} \cdot \vec{v})\vec{w} - (u \cdot \vec{w})\vec{v} + (v \cdot w)u - (\vec{v} \cdot u)w.$$

$$= (\vec{w} \cdot \vec{v})\vec{u} - (\vec{w} \cdot \vec{u})\vec{v} \qquad\qquad (\because \vec{u} \cdot \vec{v} = \vec{v} \cdot u)$$

$$= \vec{w} \times (\vec{u} \times \vec{v})$$

$$\therefore \quad \frac{d}{dt}(\vec{u} \times \vec{v}) = \vec{w} \times (\vec{u} \times \vec{v})$$

Note: $\vec{a} \times (\vec{b} \times \vec{c}) = (\vec{a} \cdot \vec{c})\vec{b} - (\vec{a} \cdot \vec{b})\vec{c}$.

(3) A particle moves along a curve $x = e^{-t}$, $y = 2\cos 3t$, $z = 2\sin 3t$ where t is the time variable. Determine its velocity and acceleration vectors and also the magnitudes of velocity and acceleration at $t = 0$.

Solution: The position vector of the particle at time t is

$$\vec{R} = x\hat{i} + y\hat{j} + z\hat{k}$$

$$= e^{-t}\hat{i} + 2\cos 3t\hat{j} + 2\sin 3t\hat{k}$$

Then, velocity $\vec{v} = \dfrac{d\vec{R}}{dt} = -e^{-t}\hat{i} - 6\sin 3t\hat{j} + 6\cos 3t\hat{k}$

$$\text{Acceleration } \vec{A} = \frac{d^2\vec{R}}{dt^2} = e^{-t}\hat{i} - 18\cos 3t\hat{j} - 18\sin 3t\hat{j}$$

At $t = 0$,
$$\vec{v} = -\hat{i} + 6\hat{k}$$

$$\vec{A} = \hat{i} - 18\hat{j}$$

Magnitude of $\vec{v}$ and $\vec{A}$.

$$|\vec{v}| = \sqrt{1^2 + 6^2} = \sqrt{17}$$

and
$$|A| = \sqrt{1^2 + 18^2} = \sqrt{325}$$

(4) A particle moves along the curve

$$\vec{R} = (t^3 - 4t)\hat{i} + (t^2 + 4t)\hat{j} + (8t^2 - 3t^3)\hat{k} \text{ where } t \text{ denotes the time. Find the}$$

magnitudes of acceleration along the tangent and normal at time $t = 2$.

Solution: The position vector of the particle at time t is

$$\vec{R} = (t^3 - 4t)\hat{i} + (t^2 + 4t)\hat{j} + (8t^2 - 3t^3)\hat{k}$$

Then

$$\vec{v} = \frac{d\vec{R}}{dt} = (3t^2 - 4)\hat{i} + (2t + 4)\hat{j} + (16t - 9t^2)\hat{k}$$

$$\vec{A} = \frac{d^2R}{dt^2} = (6t)\hat{i} + (2)\hat{j} + (16 - 18t)\hat{k}$$

At $t = 2$

$$\vec{V} = 8\hat{i} + 8\hat{j} - 4\hat{k}$$

and
$$\vec{A} = 12\hat{i} + 2\hat{j} - 20\hat{k}$$

Since the velocity is along the tangent to the curve, the component of $\vec{A}$ along the tangent

$$= \vec{A} \cdot \frac{\vec{V}}{|\vec{V}|}$$

$$= (12\hat{i} + 2\hat{j} - 20\hat{k}) \cdot \frac{(8\hat{i} + 8\hat{j} - 4\hat{k})}{\sqrt{64 + 64 + 16}}$$

$$= \frac{96 + 16 + 80}{12} = \frac{192}{12}$$

$$= 16$$

Now, the component of A along the normal

$$= |\,\bar{A} - \text{resolved part of } A \text{ along the tangent}\,|$$

$$= \left|(12\hat{i} + 2\hat{j} - 20\hat{k}) - 16\frac{(8\hat{i} + 8\hat{j} - 4\hat{k})}{12}\right|$$

$$= \frac{1}{3}|4\hat{i} - 26\hat{j} - 44\hat{k}|$$

$$= 2\sqrt{73}$$

(5) A particle moves along the curve $x = t^3 + 1$, $y = t^2$ $z = t + 5$ where t is the time. Find the components of the velocity and acceleration at time $t = 2$ in the direction of $\hat{i} + 3\hat{j} + 2\hat{k}$.

Solution: The position vector of the particle at time t is

$$\vec{R} = x\hat{i} + y\hat{j} + z\hat{k}$$

$$= (t^3 + 1)\hat{i} + t^2 + (t + 5)\hat{k}$$

Then, velocity $V = \dfrac{d\vec{R}}{dt} = 3t^2\hat{i} + 2t\hat{j} + \hat{k}$

and acceleration $\qquad A = \dfrac{d^2\vec{R}}{dt^2} = 6t\hat{i} + 2\hat{j}$.

At $t = 2$

$$\vec{V} = 12\hat{i} + 4\hat{j} + \hat{k}$$

$$\vec{A} = 12\hat{i} + 2\hat{j}$$

$\therefore$ velocity in the direction of $\hat{i} + 3\hat{j} + 2\hat{k}$

$$= \vec{V} \cdot \frac{(i + 3\hat{j} + 2\hat{k})}{\sqrt{1^2 + 3^2 + 2^2}}$$

$$= (12\hat{i} + 4\hat{j} + \hat{k}) \cdot \frac{(\hat{i} + 3\hat{j} + 2\hat{k})}{\sqrt{14}}$$

$$= \frac{12 + 12 + 2}{\sqrt{14}} = \frac{26}{\sqrt{14}}$$

Acceleration in the direction of $\hat{i} + 3\hat{j} + 2\hat{k}$ is

$$= \vec{A} \cdot \frac{(\hat{i} + 3\hat{j} + 2\hat{k})}{\sqrt{14}}$$

$$= (12\hat{i} + 2\hat{j}) \cdot = \frac{(\hat{i} + 3\hat{j} + 2\hat{k})}{\sqrt{14}}$$

$$= \frac{12 + 6}{\sqrt{14}} = \frac{18}{\sqrt{14}}$$

VECTOR OPERATOR DEL (∇)

The vector differential operator del (∇) is defined by the equation

$$\nabla = \hat{i}\frac{\partial}{\partial x} + \hat{j}\frac{\partial}{\partial y} + \hat{k}\frac{\partial}{\partial z}$$

Gradient: The vector function $\nabla\phi$ is defined as a gradient of the scalar point function ϕ and is written as "grad ϕ" or "$\nabla\phi$".

Thus,
$$\nabla\phi = \hat{i}\frac{\partial\phi}{\partial x} + \hat{j}\frac{\partial\phi}{\partial y} + \hat{k}\frac{\partial\phi}{\partial z}$$

Clearly $\qquad\qquad$ $\nabla\phi$ is a vector quantity.

Theorem: If ϕ_1 and ϕ_2 are two differentiable scalar functions of x, y, z then

(1) $\qquad\qquad$ $\nabla(\phi_1 + \phi_2) = \nabla\phi_1 + \nabla\phi_2$

(2) $\qquad\qquad$ $\nabla(\phi_1\,\phi_2) = \phi_1\nabla\phi_2 + \phi_2\nabla\phi_1$

(3) $\qquad\qquad$ $\nabla\left(\dfrac{\phi_1}{\phi_2}\right) = \dfrac{\phi_2\nabla\phi_1 - \phi_1\nabla\phi_2}{\phi_2^2}$

Examples

(1) Find $\nabla\phi$, if $\phi = \log(x^2 + y^2 + z^2)$ at $(1, 2, 3)$

Solution: $\qquad \nabla\phi = \left(\hat{i}\dfrac{\partial}{\partial x} + \hat{j}\dfrac{\partial}{\partial y} + \hat{k}\dfrac{\partial}{\partial z}\right)\left(\log(x^2 + y^2 + z^2)\right)$

$$= \hat{i}\,\frac{2x}{x^2 + y^2 + z^2} + \hat{j}\,\frac{2y}{x^2 + y^2 + z^2} + \hat{k}\,\frac{2z}{x^2 + y^2 + z^2}$$

At $(1, 2, 3)$

$$\nabla\phi = \frac{2}{14}\hat{i} + \frac{4}{14}\hat{j} + \frac{6}{14}\hat{k} = \frac{1}{7}(\hat{i} + 2\hat{j} + 3\hat{k})$$

(2) Prove that $\qquad \nabla r^n = nr^{n-2}\vec{R}$ where

$$\vec{R} = x\hat{i} + y\hat{j} + z\hat{k} \text{ and } r = |\vec{R}|$$

Proof: Given $\vec{R} = x\hat{i} + y\hat{j} + z\hat{k}$

and $\qquad\qquad$ $r = |\vec{R}| = \sqrt{x^2 + y^2 + z^2}$

$\therefore \qquad\qquad$ $r^n = (x^2 + y^2 + z^2)^{n/2}$

Now,

$$\nabla r^n = \left(\hat{i}\frac{\partial}{\partial x} + \hat{j}\frac{\partial}{\partial y} + \hat{k}\frac{\partial}{\partial z}\right)(x^2 + y^2 + z^2)^{n/2}$$

$$= \hat{i}\frac{n}{2}(x^2 + y^2 + z^2)^{\frac{n}{2}-1}2x + \hat{j}\frac{n}{2}(x^2 + y^2 + z^2)^{\frac{n}{2}-1}2y$$

$$+ k\frac{n}{2}(x^2 + y^2 + z^2)^{\frac{n}{2}-1}2z$$

$$= n(x^2 + y^2 + z^2)^{\frac{n-2}{2}}(x\hat{i} + y\hat{j} + z\hat{k})$$

$$= nr^{n-2}\vec{R}$$

$$\therefore \quad \nabla r^n = nr^{n-2}\vec{R}$$

(3) If $\phi = f(r)$ where $r^2 = x^2 + y^2 + z^2$ then prove that

$$\nabla\phi = \frac{f'(r)}{r}\vec{R}.$$

Proof: We have $\phi = f(r)$ and $r = \sqrt{x^2 + y^2 + z^2}$.

$$\nabla\phi = \left(\hat{i}\frac{\partial}{\partial x} + \hat{j}\frac{\partial}{\partial y} + \hat{k}\frac{\partial}{\partial z}\right)f(r)$$

$$= \hat{i}\frac{\partial f(r)}{\partial x}\cdot\frac{\partial r}{\partial x} + \hat{j}\frac{\partial f(r)}{\partial y}\frac{\partial r}{\partial y} + \hat{k}\frac{\partial f(r)}{\partial r}\frac{\partial r}{\partial z}$$

$$= if'(r)\frac{2x}{2\sqrt{x^2 + y^2 + z^2}} + jf'(r)\frac{2y}{2\sqrt{x^2 + y^2 + z^2}} + kf'(r)\frac{2z}{2\sqrt{x^2 + y^2 + z^2}}$$

$$= \frac{f'(r)}{\sqrt{x^2 + y^2 + z^2}}(x\hat{i} + y\hat{j} + z\hat{k})$$

$$= \frac{f'(r)}{r}\vec{R}.$$

(4) If $\vec{V_1}$ and $\vec{V_2}$ are the vectors which join the fixed points $P_1(x_1, y_1, z_1)$ and $P_2(x_2, y_2, z_2)$ to the variable point $P(x, y, z)$ then prove that

$$\nabla(\vec{V_1}\cdot\vec{V_2}) = \vec{V_1} + \vec{V_2}$$

Proof:
$$\overrightarrow{P_1P} = \vec{V_1} = (x - x_1)\hat{i} + (y - y_1)\hat{j} + (z - z_1)\hat{k}$$

$$\overrightarrow{P_2P} = \vec{V_2} = (x - x_2)\hat{i} + (y - y_2)\hat{j} + (z - z_2)\hat{k}$$

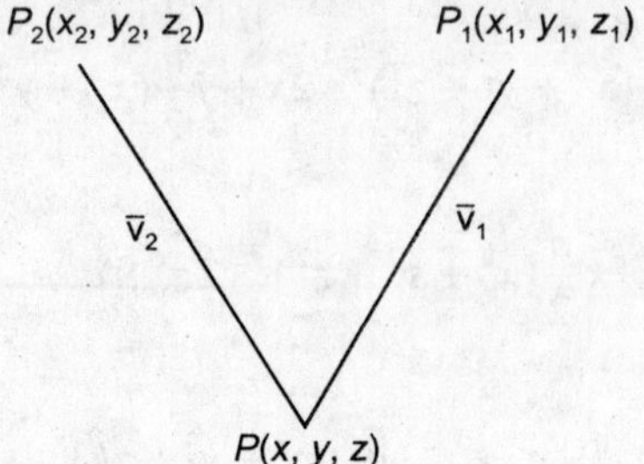

$$\therefore \quad \vec{V}_1 \cdot \vec{V}_2 = (x - x_1)(x - x_2) + (y - y_1)(y - y_2) + (z - z_1)(z - z_2)$$

$$\therefore \quad \nabla(\vec{V}_1 \cdot \vec{V}) = \left(\hat{i}\frac{\partial}{\partial x} + \hat{j}\frac{\partial}{\partial y} + \hat{k}\frac{\partial}{\partial z} \right)(\vec{V}_1 \cdot \vec{V}_2)$$

$$= \hat{i}\frac{\partial}{\partial x}(x - x_1)(x - x_2) + \hat{j}\frac{\partial}{\partial y}(y - y_1)(y - y_2)$$

$$+ \hat{k}\frac{\partial}{\partial z}(z - z_1)(z - z_2)$$

$$= \hat{i}[(x - x_1) + (x - x_2)] + \hat{j}[(y - y_1) + (y - y_2)]$$

$$+ \hat{k}[(z - z_1) + (z - z_2)]$$

$$= i(x - x_1) + j(y - y_1) + \hat{k}(z - z_1)$$

$$+ \hat{i}(x - x_2) + \hat{j}(y - y_2) + \hat{k}(z - z_2)$$

$$\therefore \quad \nabla(\vec{V}_1 \cdot \vec{V}_2) = \vec{V}_1 + \vec{V}_2$$

Definition

The grad $f((\nabla f)$ is a vector normal to the surface f = constant and has a magnitude equal to the rate of change of f along this normal.

DIRECTIONAL DERIVATIVE

The directional derivative of f in the direction of N is the resolved part of ∇f in the direction N. It follows that ∇f gives the maximum rate of change of f.

Examples

(1) Find a unit vector normal to the surface $x\,y^3\,z^2 = 4$ at the point $(-1, -1, 2)$

Solution: A vector normal to the surface

$$= \nabla\phi$$

$$= \left(\hat{i}\frac{\partial}{\partial x} + \hat{j}\frac{\partial}{\partial y} + \hat{k}\frac{\partial}{\partial z} \right)(xy^3z^2)$$

$$= \hat{i}(y^3z^2) + \hat{j}(3xy^2z^2) + \hat{k}(2xy^3z)$$

At $(-1, -1, 2)$

$$\nabla\phi = -4\hat{i} - 12\hat{j} + 4\hat{k}$$

It is a vector normal to the given surface.

$$\therefore \text{ Unit normal vector } = \frac{-4\hat{i} + 12\hat{j} + 4\hat{k}}{\sqrt{4^2 + 12^2 + 4^2}}$$

$$= \frac{-1}{\sqrt{11}}(\hat{i} + 3\hat{j} - \hat{k})$$

(2) Find a unit vector normal to the surface $x^2y + 2xz = 4$ at the point $(2, -2, 3)$.

Solution: A vector normal to the given surface is

$$\nabla\phi = \left(\hat{i}\frac{\partial}{\partial x} + \hat{j}\frac{\partial}{\partial y} + \hat{k}\frac{\partial}{\partial z} \right)(x^2y + 2xz)$$

$$= (2xy + 2z)\hat{i} + x^2\hat{j} + 2x\hat{k}$$

At $(2, -2, 3)$, $\qquad \nabla\phi = -\hat{i} + 4\hat{j} + 4\hat{k}$

It is a vector normal to the given surface.

$\therefore$ the desired unit normal vector to the surface

$$= \frac{-2\hat{i} + 4\hat{j} + 4\hat{k}}{\sqrt{4 + 16 + 16}} = \frac{1}{3}(-\hat{i} + 2\hat{j} + 2\hat{k})$$

(3) Find a unit vector normal to the surface

$$x^3 + y^3 + z^3 + 2xyz = 3 \text{ at } (1, 2, -1)$$

Solution: A vector normal to the given surface is

$$\nabla\phi = \left(\hat{i}\frac{\partial}{\partial x} + \hat{j}\frac{\partial}{\partial y} + \hat{k}\frac{\partial}{\partial z} \right)(x^3 + y^3 + z^3 + 3xyz)$$

$$= \hat{i}(3x^2 + 3yz) + \hat{j}(3y^2 + 3xz) + \hat{k}(3xy)$$

At (1, 2, –1)

$$\nabla\phi = -3\hat{i} + 9\hat{j} + 6\hat{k}$$

It is a vector normal to the given surface.

$\therefore$ Unit normal vector to the surfrace is

$$= \frac{-3\hat{i} + 9\hat{j} + 6\hat{k}}{\sqrt{9 + 81 + 36}} = \frac{3}{\sqrt{126}}(-\hat{i} + 3\hat{j} + 2\hat{k})$$

(4) Find the direction derivative of $\phi = x^2 yz + 4xz^2$ at (1, –2, –1) in the direction of $2\hat{i} - \hat{j} - 2\hat{k}$.

Solution: A vector normal to the given surfce is

$$\nabla\phi = \left(\hat{i}\frac{\partial}{\partial x} + \hat{j}\frac{\partial}{\partial y} + \hat{k}\frac{\partial}{\partial z}\right)(x^2 yz + 4xz^2)$$

$$= \hat{i}(2xyz + 4z^2) + \hat{j}(x^2 z) + \hat{k}(x^2 y + 8xz)$$

At (1, –2, –1)

$$\nabla\phi = 8\hat{i} - \hat{j} - 10\hat{k}$$

And unit vector in the direction of $2\hat{i} - \hat{j} - 2\hat{k}$ is

$$\hat{a} = \frac{2\hat{i} - \hat{j} - 2\hat{k}}{\sqrt{4 + 1 + 4}} = \frac{2}{3}\hat{i} - \frac{1}{3}\hat{j} - \frac{2}{3}\hat{k}$$

Then the directional derivative required is

$$= \nabla\phi \cdot \hat{a} = (8i - j - 10k)\cdot\left(\frac{2}{3}\hat{i} - \frac{1}{3}\hat{j} - \frac{2}{3}\hat{k}\right)$$

$$= \frac{16}{3} + \frac{1}{3} + \frac{20}{3} = \frac{37}{3}$$

Since this is positive, ϕ is increasing in this direction.

(5) Find the directional deriviative of $(x, y, z) = xy^2 + yz^3$ at the point (2, –1, 1) in the direction of vector $\hat{i} + 2\hat{j} + 2\hat{k}$.

Solution: A vector normal to the surface is

$$\nabla\phi = \left(\hat{i}\frac{\partial}{\partial x} + \hat{j}\frac{\partial}{\partial y} + \hat{k}\frac{\partial}{\partial z}\right)(xy^2 + yz^3)$$

$$= \hat{i}(y^2) + \hat{j}(2xy + z^3) + \hat{k}(3yz^2)$$

At $(2, -1, 1)$

$$\nabla\phi = \hat{i} - 3\hat{j} - 3\hat{k}$$

$\therefore$ Directional derivative of ϕ in the direction $\hat{i} + 2\hat{j} + 2\hat{k}$ is

$$= \nabla\phi \cdot \hat{a} = (\hat{i} - 3\hat{j} - 3\hat{k}) \cdot \frac{(\hat{i} + 2\hat{j} + 2\hat{k})}{\sqrt{1+4+4}}$$

$$= (i + 3j - 3k) \cdot \left(\frac{1}{3}\hat{i} + \frac{2}{3}\hat{j} + \frac{2}{3}\hat{k}\right)$$

$$= \frac{1}{3} - \frac{6}{3} - \frac{6}{3} = -\frac{11}{3}$$

Since this is negative, ϕ is decreasing in this direction.

(6) What is the directional derivative of $\phi = xy^2 + yz^3$ at the point $(2, -1, 1)$ in the direction of the normal to the surface $x \log z - y^2 = -4$ at $(-1, 2, 1)$.

Solution: A vector normal to the surface is

$$\nabla\phi = \left(\hat{i}\frac{\partial}{\partial x} + \hat{j}\frac{\partial}{\partial y} + \hat{k}\frac{\partial}{\partial z}\right)(xy^2 + yz^3)$$

$$= \hat{i}(y^2) + \hat{j}(2xy + z^3) + \hat{k}(3yz^2)$$

At $(2, -1, 1)$, $\nabla\phi = \hat{i} - 3\hat{j} - 3\hat{k}$

Again, a vector normal to the surface $x \log z - y^2 = -4$ is

$$\nabla\psi = \left(\hat{i}\frac{\partial}{\partial x} + \hat{j}\frac{\partial}{\partial y} + \hat{k}\frac{\partial}{\partial z}\right)(x\log z - y^2)$$

$$= \hat{i}(\log z) + \hat{j}(-2y) + \hat{k}\left(\frac{x}{z}\right)$$

At $(-1, 2, 1)$

$$\nabla\psi = \hat{i} - 4\hat{j} - \hat{k}$$

Directional derivative of ϕ in the direction of ∇y is

$$= \nabla \phi \cdot \frac{\nabla y}{|\nabla y|}$$

$$= (\hat{i} - 3\hat{j} - 3\hat{k}) \cdot \frac{(0\hat{i} - 4\hat{j} - \hat{k})}{\sqrt{16+1}}$$

$$= \frac{0+12+3}{\sqrt{17}} = \frac{15}{\sqrt{17}}$$

$\therefore$ ϕ is increasing in this direction.

(7) In what direction from the point $(2, 1, -1)$ is the directional derivative of $\phi\ x^2 yz^3$ a maximum? What is its magnitude?

Solution: We have

$$\nabla \phi = -4\hat{i} - 4\hat{j} + 12\hat{k} \ \text{ at } (2,\ 1,\ -1)$$

Directional derivative is maximum in the direction of $\nabla \phi$

$\therefore$ $\nabla \phi$ is maximum in the direction of $-4\hat{i} - 4\hat{j} + 12\hat{k}$ and its magnitude is

$$|\nabla \phi| = \sqrt{4^2 + 4^2 + 12} = 4\sqrt{11}$$

DIVERGENCE

Definition

The divergence of continuously differentiable vector point function $\vec{F}$ is written as $\text{div}\,\vec{F}$ or $\nabla \cdot \vec{F}$ and is defined as

$$\nabla \cdot \vec{F} = \left(\hat{i}\frac{\partial}{\partial x} + \hat{j}\frac{\partial}{\partial y} + \hat{k}\frac{\partial}{\partial z} \right) \cdot (F_1\hat{i} + F_2\hat{j} + F_3\hat{k})$$

$$= \frac{\partial F_1}{\partial x} + \frac{\partial F_2}{\partial y} + \frac{\partial F_3}{\partial z}$$

Clearly, the divergence of a vector $\vec{F}$ is a scalar quantity.

Examples

(1) If $\vec{R}$ is the position vector of any point $P(x,\ y,\ z)$ show that $\nabla \cdot \vec{R} = 3$.

Solution: Given $\vec{R} = x\hat{i} + y\hat{j} + z\hat{k}$

$$\nabla \cdot \vec{R} = \left(\hat{i}\frac{\partial}{\partial x} + \hat{j}\frac{\partial}{\partial y} + \hat{k}\frac{\partial}{\partial z} \right) \cdot (x\hat{i} + y\hat{j} + z\hat{k})$$

$$= \frac{\partial(x)}{\partial x} + \frac{\partial(y)}{\partial y} + \frac{\partial(z)}{\partial z}$$

$$= 1 + 1 + 1 = 3$$

$$\therefore \ \nabla \cdot \vec{R} = 3.$$

(2) If $\vec{A} = xy\hat{i} + yz\hat{j} + zx\hat{k},$ then find $\nabla \cdot \vec{A}$ at (1, 2, 3)

Solution: $\nabla \cdot \vec{A} = \left(\hat{i}\frac{\partial}{\partial x} + \hat{j}\frac{\partial}{\partial y} + \hat{k}\frac{\partial}{\partial z} \right) \cdot (xy\hat{i} + yz\hat{j} + 2x\hat{k})$

$$= (y + z) + (x + z) + (y + x)$$

At (1, 2, 3)

$$\nabla \cdot \vec{A} = (2 + 3) + (1 + 3) + (2 + 1)$$

$$= 5 + 4 + 3 = 12 \cdot$$

(3) If $\vec{A}$ and $\vec{B}$ are two vectors, then prove that

$$\nabla \cdot (\vec{A} + \vec{B}) = \nabla \cdot \vec{A} + \nabla \cdot \vec{B}$$

Proof: Let $\vec{A} = A_1\hat{i} + A_2\hat{j} + A_3\hat{k}$ and $\vec{B} = B_1\hat{i} + B_2\hat{j} + B_3\hat{k}$

$$\therefore \qquad \vec{A} + \vec{B} = (A_1 + B_1)\hat{i} + (A_2 + B_2)\hat{j} + (A_3 + B_3)\hat{k}$$

$$\therefore \qquad \nabla \cdot (\vec{A} + \vec{B}) = \left(i\frac{\partial}{\partial x} + j\frac{\partial}{\partial y} + k\frac{\partial}{\partial z} \right) \cdot (\vec{A} + \vec{B})$$

$$= \frac{\partial}{\partial x}(A_1 + B_1) + \frac{\partial}{\partial y}(A_2 + B_2) + \frac{\partial}{\partial z}(A_3 + B_3)$$

$$= \left(\frac{\partial A_1}{\partial x} + \frac{\partial A_2}{\partial y} + \frac{\partial A_3}{\partial z} \right) + \left(\frac{\partial B_1}{\partial x} + \frac{\partial B_2}{\partial y} + \frac{\partial B_3}{\partial z} \right)$$

$$\therefore \ \nabla \cdot (\vec{A} + \vec{B}) = \nabla \cdot \vec{A} + \nabla \cdot \vec{B}$$

(4) Show that $\nabla \cdot \nabla \phi = \nabla^2 \phi$ or div(grad ϕ) = $\nabla^2 \phi$

The operator ∇^2 is called Laplacian.

Solution: We have $\nabla \phi = \hat{i}\dfrac{\partial \phi}{\partial x} + \hat{j}\dfrac{\partial \phi}{\partial y} + \hat{k}\dfrac{\partial \phi}{\partial z}$

$$\therefore \quad \nabla \cdot \nabla \phi = \left(\hat{i}\frac{\partial}{\partial x} + \hat{j}\frac{\partial}{\partial y} + \hat{k}\frac{\partial}{\partial z} \right) \cdot \left(\hat{i}\frac{\partial \phi}{\partial x} + \hat{j}\frac{\partial \phi}{\partial y} + \hat{k}\frac{\partial \phi}{\partial z} \right)$$

$$= \frac{\partial^2 \phi}{\partial x^2} + \frac{\partial^2 \phi}{\partial y^2} + \frac{\partial^2 \phi}{\partial z^2}$$

$$= \left(\frac{\partial^2}{\partial x^2} + \frac{\partial^2}{\partial y^2} + \frac{\partial^2}{\partial z^2} \right)\phi = \nabla^2 \phi$$

THE CURL

If $v(x, y, z)$ is a differentiable vector field, then the curl or rotation of $\vec{v}$ written as $\nabla \times \vec{v}$, or curl $\vec{v}$ or rot $\vec{v}$, is defined as

$$\nabla \times \vec{v} = \begin{vmatrix} \hat{i} & \hat{j} & \hat{k} \\ \dfrac{\partial}{\partial x} & \dfrac{\partial}{\partial y} & \dfrac{\partial}{\partial z} \\ v_1 & v_2 & v_3 \end{vmatrix}$$

$$= i\left(\frac{\partial v_3}{\partial y} - \frac{\partial v_2}{\partial z} \right) - \hat{j}\left(\frac{\partial v_3}{\partial x} - \frac{\partial v_1}{\partial z} \right) + k\left(\frac{\partial v_2}{\partial x} - \frac{\partial v_1}{\partial y} \right)$$

Clearly it is a vector quantity.

Theorem: Prove that $\nabla \times (\vec{a} + \vec{b}) = \nabla \times \vec{a} + \nabla \times \vec{b}$

Proof: Let $\vec{a} = a_1\hat{i} + a_2\hat{j} + a_3\hat{k}$ and $\vec{b} = b_1\hat{i} + b_2\hat{j} + b_3\hat{k}$

$$\therefore \quad \vec{a} + \vec{b} = (a_1 + b_1)\hat{i} + (a_2 + b_2)\hat{j} + (a_3 + b_3)\hat{k}$$

$$\therefore \quad \nabla \times (\vec{a} + \vec{b}) = \begin{vmatrix} \hat{i} & \hat{j} & \hat{k} \\ \dfrac{\partial}{\partial x} & \dfrac{\partial}{\partial y} & \dfrac{\partial}{\partial z} \\ a_1 + b_1 & a_2 + b_2 & a_3 + b_3 \end{vmatrix}$$

$$= \hat{i}\left[\frac{\partial}{\partial y}(a_3+b_3) - \frac{\partial}{\partial z}(a_2+b_2)\right] - \hat{j}\left[\frac{\partial}{\partial x}(a_3+b_3) - \frac{\partial}{\partial z}(a_1+b_1)\right]$$

$$+\hat{k}\left[\frac{\partial}{\partial x}(a_2+b_2) - \frac{\partial}{\partial y}(a_1+b_1)\right]$$

$$= \hat{i}\left(\frac{\partial a_3}{\partial y} - \frac{\partial a_2}{\partial z}\right) - \hat{j}\left(\frac{\partial a_3}{\partial x} - \frac{\partial a_1}{\partial z}\right) + \hat{k}\left(\frac{\partial a_2}{\partial x} - \frac{\partial a_1}{\partial y}\right)$$

$$= \hat{i}\left(\frac{\partial b_3}{\partial y} - \frac{\partial b_2}{\partial z}\right) - \hat{j}\left(\frac{\partial b_3}{\partial x} - \frac{\partial b_1}{\partial z}\right) + \hat{k}\left(\frac{\partial b_2}{\partial x} - \frac{\partial b_1}{\partial y}\right)$$

$$= \nabla \times \vec{a} + \nabla \times \vec{b}$$

$$\therefore \ \text{curl}\ (\vec{a}+\vec{b}) = \text{curl}\ \vec{a} + \text{curl}\ \vec{b}.$$

Examples

(1) Prove that $\nabla \cdot (\phi\vec{a}) = (\nabla\phi)\cdot\vec{a} + \phi(\nabla.\vec{a})$ where ϕ is scalar or

$$\text{div}(\phi\vec{a}) = (\text{grad }\phi)\cdot\vec{a} + \phi(\text{div }\vec{a}).$$

Proof: Let $\vec{a} = a_1\hat{i} + a_2\hat{j} + a_3\hat{k}$

$$\therefore \qquad \phi\vec{a} = \phi a_1\hat{i} + \phi a_2\hat{j} + \phi a_3\hat{k}$$

$$\therefore \quad \nabla \cdot (\phi\vec{a}) = \frac{\partial}{\partial x}(\phi a_1) + \frac{\partial}{\partial y}(\phi a_2) + \frac{\partial}{\partial z}(\phi a_3)$$

$$= \left(\phi\frac{\partial a_1}{\partial x} + a_1\frac{\partial\phi}{\partial x}\right) + \left(\phi\frac{\partial a_2}{\partial y} + a_2\frac{\partial\phi}{\partial y}\right) + \left(\phi\frac{\partial a_3}{\partial z} + a_3\frac{\partial\phi}{\partial z}\right)$$

$$= \left(a_1\frac{\partial\phi}{\partial x} + a_2\frac{\partial\phi}{\partial y} + a_3\frac{\partial\phi}{\partial 2}\right) + \phi\left(\frac{\partial a_1}{\partial x} + \frac{\partial a_2}{\partial y} + \frac{\partial a_3}{\partial z}\right)$$

$$= (a_1\hat{i} + a_2\hat{j} + a_3\hat{k})\cdot\left(\frac{\partial\phi}{\partial x}\hat{i} + \frac{\partial\phi}{\partial y}\hat{j} + \frac{\partial\phi}{\partial z}\hat{k}\right)$$

$$+\phi\left(\hat{i}\frac{\partial}{\partial x} + \hat{j}\frac{\partial}{\partial y} + \hat{k}\frac{\partial}{\partial z}\right)\cdot(a_1\hat{i} + a_2\hat{j} + a_3\hat{k})$$

$$= \vec{a} \cdot \nabla \phi + \phi \nabla \cdot \vec{a}$$

$$\therefore \quad \nabla \cdot (\phi \vec{a}) = (\nabla \phi) \cdot \vec{a} + \phi (\nabla \cdot \vec{a})$$

(2) Prove that $\qquad \nabla \times (\phi \vec{a}) = (\nabla \phi) \times \vec{a} + \phi (\nabla \times \vec{a})$

or $\qquad\qquad \text{curl } (\phi \vec{a}) = (\nabla \phi) \times \vec{a} + \phi (\nabla \times \vec{a})$

Proof: Let $\vec{a} = a_1 \hat{i} + a_2 \hat{j} + a_3 \hat{k}$

$$\phi \vec{a} = \phi a_1 \hat{i} + \phi a_2 \hat{j} + \phi a_3 \hat{k}$$

$$\therefore \quad \nabla \times (\phi \vec{a}) = \begin{vmatrix} \hat{i} & \hat{j} & \hat{k} \\ \dfrac{\partial}{\partial x} & \dfrac{\partial}{\partial y} & \dfrac{\partial}{\partial z} \\ \phi a_1 & \phi a_2 & \phi a_3 \end{vmatrix}$$

$$= \hat{i} \left(\frac{\partial \phi a_3}{\partial y} - \frac{\partial \phi a_2}{\partial z} \right) - \hat{j} \left(\frac{\partial \phi a_3}{\partial x} - \frac{\partial \phi a_1}{\partial z} \right) + \hat{k} \left(\frac{\partial \phi a_2}{\partial x} - \frac{\partial \phi a_1}{\partial y} \right)$$

$$= \hat{i} \left[\left(\phi \frac{\partial a_3}{\partial y} + a_3 \frac{\partial \phi}{\partial y} \right) - \left(\phi \frac{\partial a_2}{\partial z} + a_2 \frac{\partial \phi}{\partial z} \right) \right] - \hat{j} \left[\left(\phi \frac{\partial a_3}{\partial x} + a_3 \frac{\partial \phi}{\partial x} \right) - \left(\phi \frac{\partial a_1}{\partial z} + a_1 \frac{\partial \phi}{\partial z} \right) \right]$$

$$+ \hat{k} \left[\left(\phi \frac{\partial a_2}{\partial x} + a_2 \frac{\partial \phi}{\partial x} \right) - \left(\phi \frac{\partial a_1}{\partial y} + a_1 \frac{\partial \phi}{\partial y} \right) \right]$$

$$= \hat{i} \left(a_3 \frac{\partial \phi}{\partial y} - a_2 \frac{\partial \phi}{\partial z} \right) - \hat{j} \left(a_3 \frac{\partial \phi}{\partial x} - a_1 \frac{\partial \phi}{\partial z} \right) + \hat{k} \left(a_2 \frac{\partial \phi}{\partial x} - a_1 \frac{\partial \phi}{\partial y} \right)$$

$$+ \phi \left[\hat{i} \left(\frac{\partial a_3}{\partial y} - \frac{\partial a_2}{\partial z} \right) - \hat{j} \left(\frac{\partial a_3}{\partial x} - \frac{\partial a_1}{\partial z} \right) + \hat{k} \left(\frac{\partial a_2}{\partial x} - \frac{\partial a_1}{\partial y} \right) \right]$$

$$= \begin{vmatrix} \hat{i} & \hat{j} & \hat{k} \\ \dfrac{\partial \phi}{\partial x} & \dfrac{\partial \phi}{\partial y} & \dfrac{\partial \phi}{\partial z} \\ a_1 & a_2 & a_3 \end{vmatrix} + \phi \begin{vmatrix} \hat{i} & \hat{j} & \hat{k} \\ \dfrac{\partial}{\partial x} & \dfrac{\partial}{\partial y} & \dfrac{\partial}{\partial z} \\ a_1 & a_2 & a_3 \end{vmatrix}$$

$$\therefore \quad \nabla \times (\phi \vec{a}) = \nabla \phi \times \vec{a} + \phi (\nabla \times \vec{a}) .$$

(3) Prove that $\nabla \cdot (\vec{a} \times \vec{b}) = \vec{b} \cdot (\nabla \times \vec{a}) - \vec{a} \cdot (\nabla \times \vec{b})$

or   div $(\vec{a} \times \vec{b}) = \vec{b} \cdot (\text{curl } \vec{a}) - \vec{a} \cdot (\text{curl } \vec{b})$

Proof: Let $\vec{a} = a_1 \hat{i} + a_2 \hat{j} + a_3 \hat{k}$ and $\vec{b} = b_1 \hat{i} + b_2 \hat{j} + b_3 \hat{k}$.

$$\therefore \quad \vec{a} \times \vec{b} = \begin{vmatrix} \hat{i} & \hat{j} & \hat{k} \\ a_1 & a_2 & a_3 \\ b_1 & b_2 & b_3 \end{vmatrix}$$

$$= \hat{i}(a_2 b_3 - b_2 a_3) - \hat{j}(a_1 b_3 - b_1 a_3) + \hat{k}(a_1 b_2 - b_1 a_2)$$

$$\nabla \cdot (\vec{a} \times \vec{b}) = \frac{\partial}{\partial x}(a_2 b_3 - b_2 a_3) - \frac{\partial}{\partial y}(a_1 b_3 - b_1 a_3) + \frac{\partial}{\partial z}(a_1 b_2 - b_1 a_2)$$

$$= \left[\left(a_2 \frac{\partial b_3}{\partial x} + b_3 \frac{\partial a_2}{\partial x} \right) - \left(a_3 \frac{\partial b_2}{\partial x} + b_2 \frac{\partial a_3}{\partial x} \right) \right]$$

$$- \left[\left(a_1 \frac{\partial b_3}{\partial y} + b_3 \frac{\partial a_1}{\partial y} \right) - \left(a_3 \frac{\partial b_1}{\partial y} + b_1 \frac{\partial a_3}{\partial y} \right) \right]$$

$$+ \left[\left(a_1 \frac{\partial b_2}{\partial z} + b_2 \frac{\partial a_1}{\partial z} \right) - \left(a_2 \frac{\partial b_1}{\partial z} + b_1 \frac{\partial a_2}{\partial z} \right) \right]$$

$$= (b_1 \hat{i} + b_2 \hat{j} + b_3 \hat{k}) \cdot \left[\hat{i} \left(\frac{\partial a_3}{\partial y} - \frac{\partial a_2}{\partial z} \right) - \hat{j} \left(\frac{\partial a_3}{\partial x} - \frac{\partial a_1}{\partial z} \right) + \hat{k} \left(\frac{\partial a_2}{\partial x} - \frac{\partial a_1}{\partial y} \right) \right]$$

$$- (a_1 \hat{i} + a_2 \hat{j} + b_3 \hat{k}) \cdot \left[\hat{i} \left(\frac{\partial b_3}{\partial y} - \frac{\partial b_2}{\partial z} \right) - \hat{j} \left(\frac{\partial b_3}{\partial x} - \frac{\partial b_1}{\partial z} \right) + \hat{k} \left(\frac{\partial b_2}{\partial x} - \frac{\partial b_1}{\partial y} \right) \right]$$

$$= \vec{b} \cdot (\nabla \times \vec{a}) - \vec{a} \cdot (\nabla \times \vec{b})$$

$\therefore$ div $(\vec{a} \times \vec{b}) = \vec{b} \cdot (\text{curl } \vec{a}) - \vec{a} \cdot (\text{curl } \vec{b})$

(4) Prove that $\nabla \times \nabla \phi = 0$ or curl (grad ϕ) = 0

Proof: We have $\nabla \phi = \hat{i} \dfrac{\partial \phi}{\partial x} + \hat{j} \dfrac{\partial \phi}{\partial y} + \hat{k} \dfrac{\partial \phi}{\partial z}$

$$\therefore \qquad \nabla \times \nabla \phi = \begin{vmatrix} \hat{i} & \hat{j} & \hat{k} \\ \dfrac{\partial}{\partial x} & \dfrac{\partial}{\partial y} & \dfrac{\partial}{\partial z} \\ \dfrac{\partial \phi}{\partial x} & \dfrac{\partial \phi}{\partial y} & \dfrac{\partial \phi}{\partial z} \end{vmatrix}$$

$$= \hat{i}\left(\frac{\partial^2 \phi}{\partial y \partial z} - \frac{\partial^2 \phi}{\partial z \partial y}\right) - \hat{j}\left(\frac{\partial^2 \phi}{\partial x \partial z} - \frac{\partial^2 \phi}{\partial z \partial x}\right) + \hat{k}\left(\frac{\partial^2 \phi}{\partial x \partial y} - \frac{\partial^2 \phi}{\partial y \partial x}\right)$$

$$= 0$$

(5) Prove that div (curl $\vec{A}$) = 0

or
$$\nabla \cdot (\nabla \times \vec{A}) = 0$$

Proof: Let $\vec{A} = A_1 \hat{i} + A_2 \hat{j} + A_3 \hat{k}$

$$\nabla \times \vec{A} = \begin{vmatrix} \hat{i} & \hat{j} & \hat{k} \\ \dfrac{\partial}{\partial x} & \dfrac{\partial}{\partial y} & \dfrac{\partial}{\partial z} \\ A_1 & A_2 & A_3 \end{vmatrix}$$

$$= \hat{i}\left(\frac{\partial A_3}{\partial y} - \frac{\partial A_2}{\partial z}\right) - \hat{j}\left(\frac{\partial A_3}{\partial x} - \frac{\partial A_1}{\partial z}\right) + \hat{k}\left(\frac{\partial A_2}{\partial x} - \frac{\partial A_1}{\partial y}\right)$$

$$\therefore \quad \nabla \cdot (\nabla \times \vec{A}) = \frac{\partial}{\partial x}\left(\frac{\partial A_3}{\partial y} - \frac{\partial A_2}{\partial z}\right) - \frac{\partial}{\partial y}\left(\frac{\partial A_3}{\partial x} - \frac{\partial A_1}{\partial z}\right)$$

$$+ \frac{\partial}{\partial z}\left(\frac{\partial A_2}{\partial x} - \frac{\partial A_1}{\partial y}\right)$$

$$= \frac{\partial^2 A_3}{\partial x \partial y} - \frac{\partial^2 A_2}{\partial x \partial z} - \frac{\partial^2 A_3}{\partial y \partial x} + \frac{\partial^2 A_1}{\partial y \partial z} + \frac{\partial^2 A_2}{\partial z \partial x} - \frac{\partial^2 A_1}{\partial z \partial y}$$

(6) Prove that $\nabla \times (\nabla \times \vec{A}) = \nabla(\nabla \cdot \vec{A}) - \nabla^2 \vec{A}$

Proof: We know that

$$\vec{a} \times (\vec{b} \times \vec{c}) = (\vec{a} \cdot \vec{c})\vec{b} - (\vec{a} \cdot \vec{b})\vec{c}$$

$$\therefore \quad \nabla \times (\nabla \times \vec{A}) = (\nabla \cdot \vec{A})\nabla - (\nabla \cdot \nabla)\vec{A}$$

$$= \nabla(\nabla \cdot \vec{A}) - \nabla^2 \vec{A}$$

(7) Find div $\vec{F}$ and curl $\vec{F}$, where $\vec{F} = \text{grad}(x^3 + y^3 + z^3 - 3xyz)$

Solution:

$$\vec{F} = \text{grad}(x^3 + y^3 + z^3 - 3xyz)$$

$$= \nabla(x^3 + y^3 + z^3 - 3xyz)$$

$$= \hat{i}(3x^2 - 3yz) + \hat{j}(3y^2 - 3xz) + \hat{k}(3z^2 - 3xy)$$

(i) $\nabla \cdot \vec{F} = \left(\hat{i}\dfrac{\partial}{\partial x} + \hat{j}\dfrac{\partial}{\partial y} + \hat{k}\dfrac{\partial}{\partial z} \right) \cdot \left[\hat{i}(3x^2 - 3yz) + \hat{j}(3y^2 - 3xz) + \hat{k}(3z^2 - 3xy) \right]$

$$= \frac{\partial}{\partial x}(3x^2 - 3yz) + \frac{\partial}{\partial y}(3y^2 - 3xz) + \frac{\partial}{\partial z}(3z^2 - 3xy)$$

$$= 6x + 6y + 6z = 6(x + y + z)$$

(ii) $\nabla \times \vec{F} = \begin{vmatrix} \hat{i} & \hat{j} & \hat{k} \\[4pt] \dfrac{\partial}{\partial x} & \dfrac{\partial}{\partial y} & \dfrac{\partial}{\partial z} \\[6pt] 3x^2 - 3yz & 3y^2 - 3xz & 3z^2 - 3xz \end{vmatrix}$

$$= \hat{i}(-3x + 3x)\,\hat{j}(= -3y + 3y) + \hat{k}(-3z + 3z)$$

$$= 0$$

$$\therefore \quad \nabla \cdot \vec{F} = 6(x + y + z) \text{ and } \nabla \times \vec{F} = 0.$$

(8) If $\vec{r}$ is the PV of $P(x, y, z)$ and $r = \sqrt{x^2 + y^2 + z^2}$ then show that $\nabla \cdot \left(\dfrac{\vec{r}}{r^3} \right) = 0$

Solution: $\nabla \cdot \left(\dfrac{\vec{r}}{r^3} \right) = \nabla \cdot (r^{-3}\vec{r})$

$$= (\nabla r^{-3}) \cdot \vec{r} + r^{-3}(\nabla \cdot \vec{r})$$

$$= -3\vec{r}^{5}\vec{r} \cdot \vec{r} + r^{-3}$$

$$= -3r^{-3} + 3r^{-3} \quad (\because \vec{r} \cdot \vec{r} = r^2)$$

$$= 0$$

Note the following results.

(i) $\vec{R} \cdot \vec{R} = r^2$

(ii) $\vec{R} \times \vec{R} = 0$

(iii) $\nabla \cdot \vec{R} = 3$

(iv) $\nabla \times \vec{R} = 0$

(v) $\nabla r^n = nr^{n-2}\vec{R}$.

(9) Prove that $\nabla \cdot (r^n \vec{R}) = (n+3)r^n$

Proof: $\nabla \cdot (r^n \vec{R}) = (\nabla r^n) \cdot \vec{R} + r^n (\nabla \cdot \vec{R})$

$$= nr^{n-2}\vec{R} \cdot \vec{R} + r^n (3)$$

$$= nr^{n-2}r^2 + 3r^n$$

$$= (n+3)r^n$$

(10) Prove that $\nabla \times (r^n \vec{R}) = 0$

Proof: $\nabla \times (r^n \vec{R}) = (\nabla r^n) \times \vec{R} + r^n (\nabla \times \vec{R})$

$$= nr^{n-2}\vec{R} \times \vec{R} + r^n (\nabla \times \vec{R})$$

$$= 0 + 0 = 0$$

(11) Prove that $\nabla^2 r^n = n(n+1)r^{n-2}$

Proof: $\nabla^2 r^n = \nabla \cdot \nabla r^n = \nabla \cdot (nr^{n-2}\vec{R})$

$$= n\nabla \cdot (r^{n-2}\vec{R})$$

$$= n[\nabla r^{n-2} \cdot \vec{R} + r^{n-2}\nabla \cdot \vec{R}]$$

$$= n[(n-2)r^{n-4}\vec{R} \cdot \vec{R} + 3r^{n-2}] \qquad \because \nabla \cdot \vec{R} = 3$$

$$= n[(n-2)r^{n-4}r^2 + 3r^{n-2}]$$

$$= n[(n-2+3)]r^{n-2}$$

$$= n(n+1)r^{n-2}$$

(12) If $\vec{v} = \overline{w} \times \vec{R}$ where $\overline{w}$ is a constant vector and $\vec{R} = x\hat{i} + y\hat{j} + z\hat{k}$, show that

$$\vec{w} = \frac{1}{2}\ \text{curl}\ \vec{V}$$

Solution: $\vec{R} = x\hat{i} + y\hat{j} + z\hat{k}$ and $\vec{w} = w_1\hat{i} + w_2\hat{j} + w_3\hat{k}$.

Then $\qquad \vec{v} = \overline{w} \times \vec{R}$

$$= \begin{vmatrix} \hat{i} & \hat{j} & \hat{k} \\ w_1 & w_2 & w_3 \\ x & y & z \end{vmatrix}$$

$$= \hat{i}(w_2 z - w_3 y) - \hat{j}(w_1 z - w_3 x) + \hat{k}(w_1 y - w_2 x)$$

$$\therefore \quad \text{curl}\ \vec{v} = \begin{vmatrix} \hat{i} & \hat{j} & \hat{k} \\ \dfrac{\partial}{\partial x} & \dfrac{\partial}{\partial y} & \dfrac{\partial}{\partial z} \\ (w_2 z - w_3 y) & -(w_1 z - w_3 x) & (w_1 y - w_2 x) \end{vmatrix}$$

$$= \hat{i}(w_1 + w_2) - \hat{j}(-w_2 - w_2) + \hat{k}(w_3 + w_3)$$

$$= 2(w_1\hat{i} + w_2\hat{j} + w_3\hat{k}) = 2\vec{w}$$

$$\therefore \quad \vec{w} = \frac{1}{2}\text{curl}\ \vec{v}$$

(13) Prove that $\nabla \times (\vec{a} \times \vec{r}) = 2\vec{a}$ where $\vec{a}$ is a constant vector.

Proof: Let $\vec{a} = a_1 i + a_2 j + a_3 k$ and $\vec{r} = x\hat{i} + y\hat{j} + z\hat{k}$

$$\therefore \quad \vec{a} \times \vec{r} = \begin{vmatrix} \hat{i} & \hat{j} & \hat{k} \\ a_1 & a_2 & a_3 \\ x & y & z \end{vmatrix} = \hat{i}(a_2 z - a_3 y) - \hat{j}(a_1 z - a_3 x) + \hat{k}(a_1 y - a_2 x)$$

$$\therefore \quad \nabla \times (\vec{a} \times \vec{r}) = \begin{vmatrix} \hat{i} & \hat{j} & \hat{k} \\ \dfrac{\partial}{\partial x} & \dfrac{\partial}{\partial y} & \dfrac{\partial}{\partial z} \\ a_2 z - a_3 y & -(a_1 z - a_3 x) & (a_1 y - a_2 x) \end{vmatrix}$$

$$= \hat{i}(a_1 + a_1) - \hat{j}(-a_2 - a_2) + \hat{k}(a_3 + a_3)$$

$$= 2a_1\hat{i} + 2a_2\hat{j} + 2a_3\hat{k} = 2\vec{a}$$

$$\therefore \quad \nabla \times (\vec{a} \times \vec{r}) = 2\vec{a}$$

Irrotational

Any motion in which curl of the velocity vector is not zero is said to be rotational. Otherwise it is irrotational, i.e., a vector $\vec{v}$ is irrotational if $\nabla \times \vec{v} = 0$

Solenoidal

If the flux entering any element of the space is the same as that leaving it, i.e., $\operatorname{div} \vec{v} = 0$ everywhere then such a point function is called a solenoidal vector function.

i.e., $\nabla \cdot (\nabla \phi) = 0$. Then $\nabla \phi$ is solenoidal.

Conservative

Every irrotational vector field is conservative

i.e., $\nabla \times \vec{v} = 0$.

Angle between two Surfaces

Angle between two surfaces at a point is the angle between the normals to the two surfaces at that point.

If $\nabla \phi_1$ and $\nabla \phi_2$ are the normals to the two surfaces ϕ_1 and ϕ_2 respectively then

$$\cos \theta = \frac{\nabla \phi_1 \cdot \nabla \phi_2}{|\nabla \phi_1||\nabla \phi_2|}$$

Examples

(1) Show that the vector field $\vec{F} = x^2\hat{i} + y^2\hat{j} + z^2\hat{k}$ is conservative.

Solution: We have to prove that $\nabla \times \vec{F} = 0$

$$\text{Consider,} \quad \Delta \times \vec{F} = \begin{vmatrix} \hat{i} & \hat{j} & \hat{k} \\ \dfrac{\partial}{\partial x} & \dfrac{\partial}{\partial y} & \dfrac{\partial}{\partial z} \\ x^2 & y^2 & z^2 \end{vmatrix}$$

$$= \hat{i}\left(\frac{\partial}{\partial y}(z^2) - \frac{\partial}{\partial z}(y^2)\right) - \hat{j}\left(\frac{\partial}{\partial x}(z^2) - \frac{\partial}{\partial z}(x^2)\right) + \hat{k}\left(\frac{\partial}{\partial x}(y^2) - \frac{\partial}{\partial y}(x^2)\right)$$

$$= \hat{i}(0) - \hat{j}(0) + \hat{k}(0)$$

$$= 0 \qquad\qquad\qquad \therefore\ \vec{F}\ \text{is conservative.}$$

(2) Show that the vector field $\vec{F} = yz\hat{i} + zx\hat{j} + xy\hat{k}$ is irrotational and solenoidal.

Solution: $\vec{F} = yz\hat{i} + zx\hat{j} + xy\hat{k}$

(i) $\quad \nabla \times \vec{F} = \begin{vmatrix} \hat{i} & \hat{j} & \hat{k} \\ \dfrac{\partial}{\partial x} & \dfrac{\partial}{\partial y} & \dfrac{\partial}{\partial z} \\ yz & zx & xy \end{vmatrix}$

$$= \hat{i}(x - x) - \hat{j}(y - y) + \hat{k}(z - z)$$

$$= 0$$

$\therefore\ \vec{F}$ is irrotational vector field.

(ii) $\quad \nabla \cdot \vec{F} = \left(\hat{i}\dfrac{\partial}{\partial x} + \hat{j}\dfrac{\partial}{\partial y} + \hat{k}\dfrac{\partial}{\partial x} \right) \cdot (yz\hat{i} + zx\hat{j} + xy\hat{k})$

$$= \frac{\partial}{\partial x}(yz) + \frac{\partial}{\partial y}(zx) + \frac{\partial}{\partial z}(xy)$$

$$= 0 + 0 + 0 = 0$$

$\therefore\ \vec{F}$ is solenoidal vector field.

(3) If $\vec{A}$ and $\bar{B}$ are irrotational show that $\vec{A} \times \bar{B}$ is solenoidal.

Solution: Given $\nabla \times \vec{A} = 0$ and $\nabla \times \vec{B} = 0$

We have to prove that $\nabla \cdot (\vec{A} \times \vec{B}) = 0$

$$\nabla \cdot (\vec{A} \times \vec{B}) = \vec{B} \cdot (\nabla \times \vec{A}) - \vec{A} \cdot (\nabla \times \vec{B})$$

$$= \vec{B} \cdot (0) - \vec{A} \cdot (0)$$

$$= 0$$

$\therefore \vec{A} \times \vec{B}$ is solenoidal vector field.

(4) Find the angle between the surfaces $x^2 + y^2 + z^2 = 9$ and $z = x^2 + y^2 - 3$ at $(2, -1, 2)$

Solution: Angle between two surfaces at a point is the angle between the normals to the surfaces at that point.

Let $\phi_1 = x^2 + y^2 + z^2 - 9$ and $\phi_2 = x^2 + y^2 - z - 3$ be the two surfaces. Then

$$\nabla\phi_1 = 2x\hat{i} + 2y\hat{j} + 2z\hat{k} \text{ and } \nabla\phi_2 = 2x\hat{i} + 2y\hat{j} + \hat{k}$$

$At(2, -1, 2)$

$$\nabla\phi_1 = 4\hat{i} - 2\hat{j} + 4\hat{k} \text{ and } \nabla\phi_2 = 4\hat{i} - 2\hat{j} - \hat{k}.$$

$\nabla\phi_1$ and $\nabla\phi_2$ are along the normals at $(2, -1, 2)$ If θ be the angle between these vectors which are along the normals then

$$\cos\theta = \frac{\nabla\phi_1 \cdot \nabla\phi_2}{|\nabla\phi_1||\nabla\phi_2|}$$

$$= \frac{(4\hat{i} - 2\hat{j} + 4\hat{k}) \cdot (4\hat{i} - 2\hat{j} - \hat{k})}{\sqrt{4^2 + 2^2 + 4^2}\sqrt{4^2 + 2^2 + 1^2}}$$

$$= \frac{16 + 4 - 4}{6\sqrt{21}} = \frac{16}{6\sqrt{21}}$$

(5) Find the angle between the normals to the surface $xy = z^2$ at $(1, 4, 1)$ and $(-3, -3, 3)$.

Solution: Let $\phi = xy - z^2$ be the surface.

$$\therefore \qquad \nabla\phi = \left(\hat{i}\frac{\partial}{\partial x} + \hat{j}\frac{\partial}{\partial y} + \hat{k}\frac{\partial}{\partial z} \right)(xy - z^2)$$

$$= y\hat{i} + x\hat{j} - 2z\hat{k}$$

At $(1, 4, 1)$

$$\nabla\phi_1 = 4\hat{i} + 1\hat{j} - 2\hat{k}$$

At $(-3, -3, 3)$

$$\nabla\phi_2 = -3\hat{i} - 3\hat{j} - 6\hat{k}$$

If θ be angle between $\nabla\phi_1$ and $\nabla\phi_2$ then

$$\cos\theta = \frac{\nabla\phi_1 \cdot \nabla\phi_2}{|\nabla\phi_1||\nabla\phi_2|}$$

$$= \frac{(4i + j - 2k)\cdot(-3i - 3j - 6k)}{\sqrt{16+1+4}\ \sqrt{9+9+36}}$$

$$= \frac{-12-3+12}{\sqrt{21}\cdot 3\sqrt{6}} = \frac{-1}{\sqrt{21\times 6}} = \frac{-1}{3\sqrt{14}}$$

$$\therefore\ \theta = \cos^{-1}\left(\frac{-1}{3\sqrt{14}}\right).$$

(6) Find the angle between the normals to the surfaces $x^2y + z = 3$ and $x \log z - y^2 = 4$ at point $(-1, 2, 1)$.

Solution: Let $\phi_1 = x^2y + z - 3$ and $\phi_2 = x \log z - y^2 - 4$ be the two surfaces.

$$\therefore\ \nabla\phi_1 = 2xy\hat{i} + x^2\hat{j} + \hat{k} \text{ and } \nabla\phi_2 = \log z\,\hat{i} - 2y\hat{j} + x\hat{k}$$

At $(-1, 2, 1)$

$$\nabla\phi_1 = -4\hat{i} + \hat{j} + \hat{k} \text{ and } \nabla\phi_2 = 0\hat{i} - 4\hat{j} - \hat{k}$$

If θ be the angle between these normals then

$$\cos\theta = \frac{\nabla\phi_1 \cdot \nabla\phi_2}{|\nabla\phi_1||\nabla\phi_2|}$$

$$= \frac{(-4\hat{i} + \hat{j} + \hat{k})\cdot(0\hat{i} - 4\hat{j} - \hat{k})}{\sqrt{16+1+1}\sqrt{16+1}}$$

$$= \frac{0-4-1}{3\sqrt{2}\sqrt{17}} = \frac{-5}{3\sqrt{34}}$$

(7) A vector field is given by $\vec{F} = (x^2 - y^2 + x)\hat{i} - (2xy + y)\hat{j}$ show that the field is irrotational and find its scalar potential.

Solution: We have to show that $\nabla\times\vec{F} = 0$

$$\therefore\ = \nabla\times\vec{F} = \begin{vmatrix} \hat{i} & \hat{j} & \hat{k} \\ \dfrac{\partial}{\partial x} & \dfrac{\partial}{\partial y} & \dfrac{\partial}{\partial z} \\ x^2 - y^2 + x & -(2xy + y) & 0 \end{vmatrix}$$

$$= \hat{i}(0+0) - \hat{j}(0-0) + \hat{k}(-2y+2y)$$

$$= 0$$

$\therefore$ $\nabla \times \vec{F} = 0$. Hence, the field is irrotational and the vector $\vec{F}$ can be expressed as the gradient of a scalar potential $\vec{F} = \nabla \phi$.

$\therefore$
$$(x^2 - y^2 + x)\hat{i} - (2xy + y)\hat{j} = \hat{i}\frac{\partial \phi}{\partial x} + \hat{j}\frac{\partial \phi}{\partial y}$$

$\therefore$
$$\frac{\partial \phi}{\partial x} = x^2 - y^2 + x \tag{1}$$

$$\frac{\partial \phi}{\partial y} = -2xy - y \tag{2}$$

Integrate (1) w.r.t. x, keeping y constant.

$\therefore$
$$\phi = \frac{x^3}{3} - y^2 x + \frac{x^2}{2} + f(y) \tag{3}$$

Integrate (2) w.r.t. y, keeping x constant

$\therefore$
$$\phi = -xy^2 - \frac{y^2}{2} + 9(x) \tag{4}$$

Equating (3) and (4)

$$f(y) = -\frac{y^2}{2} \text{ and } g(x) = \frac{x^3}{3} + \frac{x^2}{2}$$

$\therefore$
$$\phi = \frac{x^3}{3} - xy^2 + \frac{x^2}{2} - \frac{y^2}{2}$$

(8) Show that the vector field

$$\vec{F} = (x^2 - yz)\hat{i} + (y^2 - zx)\hat{j} + (z^2 - xy)\hat{k} \text{ is irrotational}$$

and find its scalar potential

Solution: We have to show that $\nabla \times \vec{F} = 0$

$\therefore$
$$\nabla \times \vec{F} = \begin{vmatrix} \hat{i} & \hat{j} & \hat{k} \\ \dfrac{\partial}{\partial x} & \dfrac{\partial}{\partial y} & \dfrac{\partial}{\partial z} \\ x^2 - yz & y^2 - zx & z^2 - xy \end{vmatrix}$$

$$= \hat{i}(-x+x) - \hat{j}(-y+y) + \hat{k}(-z+x)$$

$$= 0$$

$\therefore$ The field is irrotational and the vector $\vec{F}$ can be expressed as the gradient of a scalar potential.

$\therefore$
$$\vec{F} = \nabla \phi$$

$$(x^2 - yz)\hat{i} + (y^2 - zx)\hat{j} + (z^2 - xy)\hat{k} = \hat{i}\frac{\partial \phi}{\partial x} + \hat{j}\frac{\partial \phi}{\partial y} + \hat{k}\frac{\partial \phi}{\partial z}$$

$\therefore$
$$\frac{\partial \phi}{\partial x} = x^2 - yz; \quad \frac{\partial \phi}{\partial y} = y^2 - zx; \quad \frac{\partial \phi}{\partial z} = z^2 - xy$$

Consider

$$d\phi = \frac{\partial \phi}{\partial x}dx + \frac{\partial \phi}{\partial y}dy + \frac{\partial \phi}{\partial z}dz$$

$$= (x^2 - yz)dx + (y^2 - 2x)dy + (z^2 - xy)dz$$

Integrating

$$\phi = \underset{\substack{\text{keeping} \\ y\,\&\,z \\ \text{constant}}}{\int (x^2 - yz)dx} + \underset{\substack{\text{ignore } x \\ \&\text{keep} \\ z \text{ constant}}}{\int y^2 dy} + \underset{\substack{\text{ignore} \\ x\,\&\,y}}{\int z^2 dz} + c_1$$

$$= \frac{x^3}{3} - xyz + \frac{y^2}{3} + \frac{z^2}{3} + c_1$$

$$\phi = x^3 + y^3 + z^3 - 3xyz + c \quad \text{where } 3c_1 = c.$$

(9) Find the constants a and b so that

$$\vec{F} = (axy + z^3)\hat{i} + (3x^2 - z)\hat{j} + (bxz^2 - y)\hat{k} \quad \text{is irrotational}$$

and find ϕ such that $\vec{F} = \nabla \phi$.

Solution: We have to prove that $\nabla \times \vec{F} = 0$

$$\therefore \quad \nabla \times \vec{F} = \begin{vmatrix} \hat{i} & \hat{j} & \hat{k} \\ \dfrac{\partial}{\partial x} & \dfrac{\partial}{\partial y} & \dfrac{\partial}{\partial z} \\ axy + z^3 & 3x^2 - z & bxz^2 - y \end{vmatrix} = 0$$

$$\therefore \qquad \hat{i}(-1+1) - \hat{j}(bz^2 - 2z^2) + \hat{k}(6x - ax) = 0$$

This will be zero if $b - 3 = 0$ and $6 - a = 0$

$$\therefore \quad b = 3 \text{ and } a = 6$$

Now
$$\nabla\phi = \vec{F}$$

$$\therefore \qquad \hat{i}\frac{\partial\phi}{\partial x} + \hat{j}\frac{\partial\phi}{\partial y} + \hat{k}\frac{\partial\phi}{\partial z} = (6xy + z^3)\hat{i} + (3x^2 - z)\hat{j} + (3xz^2 - y)\hat{k}$$

$$\therefore \qquad \frac{\partial\phi}{\partial x} = 6xy + z^3, \frac{\partial\phi}{\partial y} = 3x^2 - z \text{ and } \frac{\partial\phi}{\partial z} = 3xz^2 - y$$

$$\therefore \qquad d\phi = (6xy + z^3)dx + (3x^2 - z)dy + (3xz^2 - y)dz$$

Integrating

$$\phi = \int\limits_{\substack{y\&z \\ \text{constant}}} (6xy + z^3)dx + \int\limits_{\substack{\text{ignore} \\ z\&x \\ \text{constant}}} (-z)dy + \int\limits_{\substack{\text{ignore} \\ \text{both} \\ x\&y}} 0\,dz + c$$

$$\phi = 6y\frac{x^2}{2} + z^3 x + (-zy) + c$$

$$\phi = 3x^2 y + xz^3 - yz + c.$$

(10) If $\vec{V} = 2xy^2\,\hat{i} + 3x^2 y\,\hat{j} - 3ayz\,\hat{k}$ is solenoidal at $(1, 1, 1)$, find a.

Solution: We have $\nabla \cdot \vec{V} = 0$

$$\left(\hat{i}\frac{\partial}{\partial x} + \hat{j}\frac{\partial}{\partial y} + \hat{k}\frac{\partial}{\partial z}\right) \cdot (2xy\hat{i} + 3x^2 y\hat{j} - 3ayz\hat{k}) = 0$$

$$\therefore \qquad \frac{\partial}{\partial x}(2xy) + \frac{\partial}{\partial y}(3x^2 y) + \frac{\partial}{\partial z}(-3ayz) = 0$$

$$\therefore \qquad 2y + 3x^2 - 3ay = 0$$

At $(1, 1, 1)$, $2 \times 1 \ 3 \times 1 - 3a \times 1 = 0$

$$\therefore \quad 5 - 3a = 0$$

$$\therefore \quad a = 5/3$$

(11) Find the constants a, b, c so that the vector

$$\vec{F} = (x + 2y + az)\hat{i} + (bx - 3y - z)\hat{j} + (4x + cy + 2z)\hat{k}$$

is irrotational.

Solution: We have $\nabla \times \vec{F} = 0$

$$\begin{vmatrix} \hat{i} & \hat{j} & \hat{k} \\ \dfrac{\partial}{\partial x} & \dfrac{\partial}{\partial y} & \dfrac{\partial}{\partial z} \\ x + 2y + az & bx - 3y - z & 4x + cy + 2z \end{vmatrix} = 0$$

$$\therefore \qquad \hat{i}(c + 1) - \hat{j}(4 - a) + \hat{k}(b - 2) = 0$$

$$\therefore \qquad c + 1 = 0,\ 4 - a = 0,\ b - 2 = 0$$

$$\therefore \qquad a = 4,\ b = 2,\ c = -1$$

QUESTION BANK 4

(1) If $\vec{A} = 5t^2\hat{i} + t\hat{j} - t^3\hat{k}$, $\vec{B} = \sin t\hat{i} - \cos t\hat{j}$

     Find (i) $\dfrac{d\,(\vec{A} \cdot \vec{B})}{dt}$ (ii) $\dfrac{d\,(\vec{A} \cdot \vec{B})}{dt}$

(2) If $\dfrac{d\bar{A}}{dt} = \vec{w} \times \vec{A}$ and $\dfrac{d\bar{B}}{dt} = \vec{w} \times \vec{B}$ then prove that $\dfrac{d}{dt}(\bar{A} \times \bar{B}) = \vec{w} \times (\bar{A} \times \bar{B})$.

(3) Find the velocity and acceleration of a particle which moves along the curve $\vec{R} = e^{-2t}\hat{i} + (2\cos 5t)\hat{j} + (5\sin 2t)\hat{k}$. Also find the magnitudes at $t = 0$.

(4) A particle moves along the curve $x = t^3$, $y = t^2$, $z = 8t^2 - 3t^3$ where t denotes time. Find the magnitudes of velocity and acceleration in the direction of the vector $2\hat{i} + 2\hat{j} - \hat{k}$ at $t = 2$.

(5) A particle moves so that its position vector is given by $\vec{r} = \cos wt\hat{i} + \sin wt\hat{j}$ where w is constant. Show that

   (i) The velocity $\vec{v}$ of the particle is $\perp$ to $\vec{r}$.

   (ii) The acceleration $\vec{A}$ is directed towards the origin.

(6) If $\vec{r} = \sin t\,\hat{i} + \cos t\,\hat{j} + t\,\hat{k}$, then find $\dfrac{d\vec{r}}{dt}$, $\dfrac{d^2\vec{r}}{dt^2}$, $\left|\dfrac{d\vec{r}}{dt}\right|$ and $\left|\dfrac{d^2\vec{r}}{dt^2}\right|$

(7) A particle moves along the curve $\vec{r} = 3t^2\hat{i} + (t^3 - 4t)\hat{j} + (3t + 4)\hat{k}$. Find the components of velocity and acceleration at $t = 2$ in the direction of $\hat{i} = 2\hat{j} + 2\hat{k}$.

(8) Find the directional derivative of the function $\phi = xyz$, along the direction of the normal to the surface $xy^2 + yz^2 + zx^2 = 3$ at the point $(1, 1, 1)$.

(9) Find a unit vector normal to the surface $xy^3z^2 = 4$ at the point $(-1, -1, 2)$.

(10) Find the directional derivative of $\phi = x^2 - 2xy + z^3$ at the point $(2, -1, 1)$ in the direction of $2\hat{i} - 4\hat{j} + 4\hat{k}$.

(11) Find the div $\vec{F}$ and curl $\vec{F}$ where $\vec{F} = \nabla(x^3 + y^3 + z^3 - 3xyz)$.

(12) Find the constants a, b, c such that the vector field

$$(\sin y + az)\hat{i} + (bx\cos y + z)\hat{j} + (x + cy)\hat{k}$$

is irrotational. Also find the scalar potential ϕ such that $\vec{F} = \nabla\phi$.

(13) Prove that $\nabla^2(\log r) = \dfrac{1}{r^2}$ where $\vec{r} = x\hat{i} + y\hat{j} + z\hat{k}$ and $r = |\vec{r}|$.

(14) If $\vec{F} = (x + y + 1)\hat{i} + \hat{j} - (x + y)\hat{k}$ show that $\vec{F} \cdot \text{curl } \vec{F} = 0$.

(15) Prove that $\nabla \cdot (\vec{a} \times \vec{b}) = \vec{b} \cdot (\nabla \times \vec{a}) - \vec{a} \cdot (\nabla \times \vec{b})$.

(16) Find the constants a and b so that the vector

$$\vec{F} = (axy + z^3)\hat{i} + (3x^2 - z)\hat{j} + (bxz^2 - y)\hat{k} \text{ is irrotational.}$$

(17) If $\vec{V_1}$ and $\vec{V_2}$ be the vectors joining fixed points (x_1, y_1, z_1) and (x_2, y_2, z_2) respectively to a variable point then prove that div $(\vec{V_1} \times \vec{V_2}) = 0$.

(18) Find the unit tangent vector at any point on the curve $x = t^3 + 1$, $y = 4t - 3$, $z = 2t^2 - 6t$.

(19) Prove that $\nabla \times (\nabla \times \vec{F}) = \nabla(\nabla \cdot \vec{F}) - \nabla^2 F$.

(20) Show that $\vec{F} = \dfrac{xi + yj}{x^2 + y^2}$ is both solenoidal and irrotational.

(21) Find the divergence and curl of the vector

$$\vec{F} = (xyz + y^2z)\hat{i} + (3x^2y + y^2z)\hat{j} + (xz^2 - y^2z)\hat{k}$$

PART (C)
Laplace Transforms

Laplace Transforms-1

A transformation is a mathematical device which converts one function into another. Laplace transformation is a transform widely used by scientists and engineers to solve linear differential equations effectively, as well as partial differential equations. Laplace transform directly gives the solution of differential equations with given initial conditions.

Definition: The integral transform $f(t)$ is given by

$$I\{f(t)\} = \int_{-\infty}^{\infty} k(s,t) f(t) dt = F(s)$$

where $k(s,t)$ is called the KERNEL of the transformation and s is the transformation parameter.

If $k(s,t)$ is selected such that

$$\begin{aligned} k(s,t) &= 0 \quad &&\text{for} \quad t < 0 \\ &= e^{-st} \quad &&\text{for} \quad t \geq 0 \end{aligned}$$

then the integral transform is called the Laplace transform.

Definition: If $f(t)$ is a function of t, then the definite integral $\int_{0}^{\infty} e^{-st} f(t) dt$, if it exists, is a function of the parameter s and is denoted by $F(s)$ or $L\{f(t)\}$.

Thus,

$$L\{f(t)\} = F(s) = \int_{0}^{\infty} e^{-st} f(t) dt$$

where L is the Laplace transform operator.

Also, if L^{-1} is the inverse Laplace transform operator

then $L^{-1}\{F(s)\} = f(t)$.

LAPLACE TRANSFORMS OF SOME ELEMENTARY FUNCTIONS

(1) Let $f(t) = e^{at}$

By definition
$$L\{e^{at}\} = \int_0^\infty e^{-st} e^{at} dt$$

$$= \int_0^\infty e^{-(s-a)t} dt$$

$$= \left[\frac{e^{-(s-a)t}}{-(s-a)} \right]_0^\infty = 0 - \frac{1}{-s-a}$$

$$\therefore \qquad L\{e^{at}\} = \frac{1}{s-a} \tag{1}$$

Cor 1: Replace $-a$ for a in (1)

$$L\{e^{-at}\} = \frac{1}{s-(-a)} = \frac{1}{s+a}$$

$$\therefore \qquad L\{e^{-at}\} = \frac{1}{s+a} \tag{2}$$

Cor 2: Put $a = 0$ in (1)

$$L\{1\} = \frac{1}{s} \tag{3}$$

Cor 3: Let $f(t) = a^t$

$$\therefore \qquad L\{a^t\} = L\{e^{t\log a}\}$$

$$= \frac{1}{s - \log a}$$

$$L\{a^t\} = \frac{1}{s - \log a} \tag{4}$$

(2) Let $f(t) = \sin at$

By definition
$$L\{\sin at\} = \int_0^\infty e^{-st} \sin at\, dt$$

$$= \left[e^{-st} \frac{(-s\sin at - a\cos at)}{(-s)^2 + a^2} \right]_0^\infty$$

$$= 0 - \frac{(-a)}{s^2 + a^2} = \frac{a}{s^2 + a^2}$$

$$\therefore \qquad L\{\sin at\} = \frac{a}{s^2 + a^2} \qquad (5)$$

(3) Let $f(t) = \cos at$

By definition
$$L\{\cos at\} = \int_0^\infty e^{-st} \cos at \; dt$$

$$= e^{-st} \frac{(-s \cos at + a \sin at)}{s^2 + a^2} \bigg]_0^\infty$$

$$= 0 - \frac{(-s)}{s^2 + a^2} = \frac{s}{s^2 + a^2}$$

$$\therefore \qquad L\{\cos at\} = \frac{s}{s^2 + a^2} \qquad (6)$$

(4) Let
$$f(t) = \sinh at$$

$$= \frac{e^{at} - e^{-at}}{2}$$

$$\therefore \qquad L\{\sinh at\} = \frac{1}{2} L\{e^{at} - e^{-at}\}$$

$$= \frac{1}{2} \left[\frac{1}{s-a} - \frac{1}{s+a} \right]$$

$$= \frac{1}{2} \left[\frac{s+a-(s-a)}{s^2 - a^2} \right] = \frac{a}{s^2 + a^2}$$

$$\therefore \qquad L\{\sinh at\} = \frac{a}{s^2 - a^2} \qquad (7)$$

(5) Let $f(t) = \cosh at$

$$= \frac{e^{at} + e^{-at}}{2}$$

$$\therefore \qquad L\{\cosh at\} = \frac{1}{2} L\{e^{at} + e^{-at}\}$$

$$= \frac{1}{2}\left[\frac{1}{s-a} + \frac{1}{s=a}\right] = \frac{s}{s^2 - a^2}$$

$$\therefore \qquad L\{\cosh at\} = \frac{s}{s^2 - a^2} \tag{8}$$

(6) Let $f(t) = t^{n-1}$ where n is a positive integer. To find the Laplace transform of this, we use the Gamma function given by

$$\overline{\lceil n} = \int_0^\infty e^{-x} x^{n-1} dx$$

Put $\qquad\qquad x = st \quad \therefore dx = s\,dt$

$$= \int_0^\infty e^{-st} (st)^{n-1} s\,dt$$

$$= s^n \int_0^\infty e^{-st} t^{n-1} s\,dt$$

$$\overline{\lceil n} = s^n L\{t^{n-1}\} \qquad\qquad \text{by definition}$$

$$\therefore \qquad L\{t^{n-1}\} = \frac{\overline{\lceil n}}{s^n} = \frac{(n-1)!}{s^n} \qquad\qquad (\because \overline{\lceil n} = (n-1)!)$$

$$\therefore \qquad L\{t^{n-1}\} = \frac{(n-1)!}{s^n}$$

or $\qquad\qquad L\{t^n\} = \dfrac{n!}{s^{n+1}} \tag{9}$

PROPERTIES OF LAPLACE TRANSFORMS

(a) Linearity property

If a, b, c are some constants and f, g, h are functions of t, then

$$L\{af(t) + bg(t) - ch(t)\} = aL\{f(t)\} + bL\{g(t)\} - cL\{h(t)\}$$

(b) First shifting property

If $\qquad\qquad L\{f(t)\} = F(s)$ then $L\{e^{at} f(t)\} = F(s-a)$

Proof: By definition

$$L\{f(t)\} = \int_0^\infty e^{-st} f(t)\,dt$$

$\therefore$

$$L\{e^{at}(t)\} = \int_0^\infty e^{-st} e^{at} f(t)\,dt$$

$$= \int_0^\infty e^{-(s-a)t} f(t)\,dt$$

Put

$$s - a = r$$

$$= \int_0^\infty e^{-rt} f(t)\,dt$$

$$= F(r) = F(s-a) \qquad \text{(by definition)}$$

$\therefore$

$$L\{e^{-at} f(t)\} = F(s-a)$$

Cor: Replace $-a$ by a then

$$L\{e^{-at} f(t)\} = F(s+a)$$

The first shifting property leads us the following useful results.

(7)
$$L\{e^{at} t^n\} = \frac{n!}{(s-a)^{n+1}} \qquad (10)$$

(8)
$$L\{e^{at} \sin bt\} = \frac{b}{(s-a)^2 + b^2} \qquad (11)$$

(9)
$$L\{e^{at} \cos bt\} = \frac{s-a}{(s-a)^2 + b^2} \qquad (12)$$

(10)
$$L\{e^{at} \sinh bt\} = \frac{b}{(s-a)^2 - b^2} \qquad (13)$$

(11)
$$L\{e^{at} \cosh bt\} = \frac{s-a}{(s-a)^2 - b^2} \qquad (14)$$

(c) Change of scale property

If
$$L\{f(t)\} = F(s) \text{ then } L\{f(at)\} = \frac{1}{a} F\left(\frac{s}{a}\right)$$

Proof: We have

$$L\{f(at)\} = \int_0^\infty e^{-st} f(at)\, dt$$

$$= \int_0^\infty e^{-s\frac{u}{a}} f(u) \frac{du}{a} \qquad \text{by putting, } u = at$$

$$= \frac{1}{a} \int_0^\infty e^{-s\frac{u}{a}} f(u)\, du$$

$$\therefore \qquad L\{f(at)\} = \frac{1}{a} F\left(\frac{s}{a}\right)$$

Before we go to the problems, recollect some of the standard trigonometric formulae.

(1) $\sin A \cos B = \dfrac{1}{2}[\sin(A+B) + \sin(A-B)$

(2) $\sin A \sin B = \dfrac{1}{2}[\cos(B-A) - \cos(A+B)$

(3) $\cos A \cos B = \dfrac{1}{2}[\cos(A+B) + \cos(A-B)]$

(4) $\sin^2 A = \dfrac{1-\cos 2A}{2}; \ \cos^2 A = \dfrac{1+\cos 2A}{2}$

(5) $\sin^3 A = \dfrac{1}{4}(3\sin A - \sin 3A)$

(6) $\cos^3 A = \dfrac{1}{4}(\cos 3A + 3\cos A)$

(7) $\cos(A+B) = \cos A \cos B - \sin A \sin B$

(8) $\sin(A+B) = \sin A \cos B + \cos A \sin B$

(9) $\sinh ax = \dfrac{e^{ax} - e^{-ax}}{2}; \ \cosh ax = \dfrac{e^{ax} + e^{-ax}}{2}.$

Examples

Find the Laplace transforms of (1) $\sin 4t \sin 6t$ (2) $\sin^3 2t$ (3) $e^{-3t}(\cos 5t + 3\sin 5t)$

Solutions:

(1) $L\{\sin 4t \sin 6t\}$

$$= \frac{1}{2}L\{\cos(6t - 4t) - \cos(4t + 6t)\}$$

$$= \frac{1}{2}L\{\cos 2t - \cos 10t\}$$

$$= \frac{1}{2}\left[\frac{s}{s^2 + 2^2} - \frac{s}{s^2 + 10^2}\right]$$

(2) $L\{\sin^3 2t\}$

$$= L\left\{\frac{2\sin 2t - \sin 6t}{4}\right\}$$

$$= \frac{3}{4}L\{\sin 2t\} - \frac{1}{4}L\{\sin 6t\}$$

$$= \frac{3}{4} \cdot \frac{2}{s^2 + 2^2} - \frac{1}{4} \cdot \frac{6}{s^2 + 6^2}$$

$$= \frac{6}{4}\left(\frac{1}{s^2 + 4} - \frac{1}{s^2 + 36}\right) = \frac{3}{2}\left(\frac{1}{s^2 + 4} - \frac{1}{s^2 + 36}\right)$$

(3) $L\{e^{-3t}(\cos 5t + 3\sin 5t)\}$

$$= L\{e^{-3t}\cos 5t\} + 3L\{e^{-3t}\sin 5t\}$$

$$= \frac{s + 3}{(s + 3)^2 + 5^2} + 3 \cdot \frac{5}{(s + 3)^2 + 5^2}$$

$$= \frac{s + 3 + 15}{s^2 + 6s + 9 + 25}$$

$$= \frac{s + 18}{s^2 + 6s + 34}.$$

(4) If $f(t) = \begin{cases} \sin t & 0 < t < \pi \\ 0 & t > \pi \end{cases}$

Find $L\{f(t)\}$

Solution: $L\{f(t)\} = \int_0^\infty e^{-st}(f(t)dt$

$$= \int_0^\pi e^{-st}\sin t\,dt + \int_\pi^\infty e^{-st}(0)dt$$

$$= \int_0^\pi e^{-st}\sin t\,dt + 0$$

$$= e^{-st}\frac{(-s\sin t - \cos t)}{s^2 + 1^2}\Bigg]_0^\pi$$

$$= e^{-\pi s}\frac{(\cos\pi - \cos 0)}{s^2 + 1}$$

$$= \frac{-e^{-\pi s}}{s^2 + 1}(-1 - 1)$$

$$= \frac{2e^{-\pi s}}{s^2 + 1}$$

Find the Laplace transform of (5) $e^{-3t}\cos^3 2t$ (6) $e^{2t}\cos 5t \sin 2t$ (7) $e^{4t}\cosh 2t \sin 3t$.

Solutions:

(5) $L\{e^{-3t}\cos^3 2t\}$

$$= L\left\{e^{-3t}\frac{(\cos 6t + 3\cos 2t)}{4}\right\}$$

$$= \frac{1}{4}L\{e^{-3t}\cos 6t\} + \frac{3}{4}L\{e^{-3t}\cos 2t\}$$

$$= \frac{1}{4}\frac{s+3}{(s+3)^2 + 6^2} + \frac{3}{4}\frac{s+3}{(s+3)^2 + 2^2}$$

$$= \frac{1}{4}\left[\frac{s+3}{s^2 + 6s + 45} + \frac{3s+9}{s^2 + 6s + 13}\right]$$

(6) $\qquad\qquad L\{e^{2t}\cos 5t \sin 2t\}$

$$= L\left\{e^{2t}\frac{1}{2}(\sin(2t + 5t) + \sin(2t - 5t))\right\}$$

$$= \frac{1}{2} L\{e^{2t} \sin 7t\} - \frac{1}{2} L\{e^{2t} \sin 3t\}$$

$$= \frac{1}{2} \frac{7}{(s-2)^2 + 7^2} - \frac{1}{2} \frac{3}{(s-2)^2 + 3^2}$$

$$= \frac{1}{2} \left[\frac{7}{s^2 - 4s + 53} - \frac{3}{s^2 - 4s + 13} \right]$$

(7) $L\{e^{4t}(\cosh 2t \sin 3t\}$

$$= L\left\{ e^{4t} \left(\frac{e^{2t} + e^{-2t}}{2} \right) \sin 3t \right\} = \frac{1}{2} L\{e^{6t} \sin 3t + e^{2t} \sin 3t\}$$

$$= \frac{1}{2} \left[\frac{3}{(s-6)^2 + 3^2} + \frac{3}{(s-2)^2 + 3^2} \right]$$

(8) Find the Laplace transform of $f(t)$ defined as

$$f(t) = \begin{cases} t/T & \text{when} \quad 0 < t < T \\ 1 & \text{when} \quad t > T \end{cases}$$

Solution: By definition

$$L\{f(t)\} = \int_0^\infty e^{-st} f(t) dt$$

$$= \int_0^T e^{-st} \frac{t}{T} dt + \int_T^\infty e^{-st} (t) dt$$

$$= \frac{1}{T} \left[t \frac{e^{-st}}{-s} - 1 \frac{e^{-st}}{s^2} \right]_0^T + \left[\frac{e^{-st}}{-s} \right]_T^\infty$$

$$= \frac{1}{T} \left[T \frac{e^{-sT}}{-s} - \frac{e^{-sT}}{s^2} + \frac{1}{s^2} \right] - \frac{e^{-sT}}{-s}$$

$$= \frac{1}{Ts^2} (1 - e^{-sT}) = \frac{1 - e^{-st}}{Ts^2}$$

Find the Laplace transform of (9) $t\sqrt{t}$, (10) $\cos(wt + \alpha)$ (11) $e^{2t}(2t^2 + \cos 2t - \sinh 3t)$

Solutions:

(9) $L\{t\sqrt{t}\} = L\{t^{3/2}\}$

$$= \frac{\left|\frac{3}{2} + 1\right.}{s^{\frac{3}{2}+1}} = \frac{\left|\frac{5}{2}\right.}{s^{\frac{5}{2}}} = \frac{\frac{3}{2} \cdot \frac{1}{2}\left|\frac{1}{2}\right.}{s^{\frac{5}{2}}}$$

$$= \frac{3\sqrt{\pi}}{4s^{5/2}} \quad \left(\because \left|\frac{1}{2}\right. = \sqrt{\pi}\right)$$

(10) $L\{\cos(wt + \alpha)\}$

$$= L\{\cos wt \sin\alpha - \sin wt \cos\alpha\}$$

$$= \sin\alpha L\{\cos wt\} - \cos\alpha L\{\sin wt\}$$

$$= \sin\alpha \frac{s}{s^2 + w^2} - \cos\alpha \frac{w}{s^2 + w^2}$$

$$= \frac{\sin\alpha \cdot s - w\cos\alpha}{s^2 + w^2}$$

(11) $L\{e^{2t}(2t^2 + \cos 2t - \sinh 3t)\}$

$$= 2L\{e^{2t}t^2\} + L\{e^{2t}\cos 2t\} - L\{e^{2t}\sinh 3t\}$$

$$= \frac{2 \times 2!}{(s-2)^3} + \frac{s-2}{(s-2)^2 + 2^2} - \frac{3}{(s-2)^2 + 3^2}$$

$$= \frac{4}{(s-2)^3} + \frac{s-2}{s^2 - 4s + 8} - \frac{3}{s^2 - 4s + 13}$$

LAPLACE TRANSFORM OF DERIVATIVES

If $f(t)$ is of exponential order and $f'(t)$ is continuous, then,

$$L\{f'(t)\} = sF(s) - f(0).$$

Proof: By Definition

$$L\{f'(t)\} = \int\limits_0^\infty e^{-st} f'(t)dt$$

Integrating by parts

$$= [e^{-st} f(t)]_0^\infty - \int\limits_0^\infty e^{-st}(-s)f(t)dt$$

$$= 0 - f(0) + s\int\limits_0^\infty e^{-st} f(t)dt$$

$$= -f(0) + sF(s)$$

$$\therefore \qquad L\{f'(t) = sF(s) - f(0)$$

Cor: Prove that

$$L\{f''(t)\} = s^2 F(s) - sf(0) - f'(0)$$

Let $f'(t) = g(t)$ so that $f''(t) = g'(t)$

$$\therefore \quad L\{f''(t)\} = L\{g'(t)\}$$

$$= sL\{g(t)\} - g(0)$$

$$= sL\{f'(t)\} - f'(0) \qquad\qquad \because f'(0) = g(0)$$

$$= s[sL\{f(t)\} - f(0)] - f'(0)$$

$$= s^2 L\{f(t)\} - sf(0) - f'(0)$$

$$\therefore \qquad\qquad L\{f''(t)\} = s^2 F(s) - sf(0) - f'(0)$$

In general,

$$L\{f^n(t)\} = s^n F(s) - s^{n-1}f(0) - s^{n-2}f'(0) \ ... \ f^{n-1}_{(0)}$$

or If $f(t) = y(t)$ then

$$L\{y^n(t)\} = s^n L\{y(t)\} - s^{n-1}y(0) - s^{n-2}y'(0) \ ... - y^{n-1}(0)$$

LAPLACE TRANSFORM OF INTEGRAL

If $L\{f(t)\} = F(s)$ then

$$L\left\{\int\limits_0^t f(u)du\right\} = \frac{1}{s}F(s)$$

Proof: Let $\phi(t) = \int\limits_0^t f(u)\,du$

$\therefore \qquad\qquad\qquad\qquad \phi'(t) = f(t) \text{ and } \phi(0) = 0$

Now, $\qquad\qquad\qquad L\{\phi'(t)\} = sL\{\phi(t)\} - \phi(0)$

$\therefore \qquad\qquad\qquad L\{f(t)\} = sL\{\phi(t)\} - 0$

$\therefore \qquad\qquad\qquad\qquad F(s) = sL\{\phi(t)\}$

$\therefore \qquad\qquad\qquad\qquad L\{\phi(t)\} = \frac{1}{s}F(s)$

$$\therefore L\left\{ \int\limits_0^t f(u)\,du \right\} = \frac{1}{s}F(s)$$

or $\qquad\qquad L^{-1}\left\{ \frac{1}{s}F(s) \right\} = \int\limits_0^t f(u)\,du$

MULTIPLICATION BY t^n

If $\quad L\{f(t)\} = F(s)$ then prove that

$$L\{t^n f(t)\} = (-1)^n \frac{d^n}{ds^n}(F(s)), \, n = 1, 2, 3, \dots$$

Proof: By mathematical induction

Case I: $n = 1$, by definition

$$\int\limits_0^\infty e^{-st} f(t)\,dt = F(s)$$

Differentiate w.r.t. s both sides

$$\frac{d}{ds}\int\limits_0^\infty e^{-st} f(t)\,dt = \frac{d}{ds}F(s)$$

On by Leibnitz's rule differentiation under integral sign

$$\int\limits_0^\infty \frac{\partial}{\partial s}(e^{-st})f(t)\,dt = \frac{d}{ds}F(s)$$

$$\therefore \qquad \int_0^\infty e^{-st}(-t)f(t)\,dt = \frac{d}{ds}F(s)$$

$$\therefore \qquad \int_0^\infty e^{-st}(tf(t))\,dt = (-1)\frac{d}{ds}F(s)$$

Hence, the result is true for $n = 1$

Case II: Let us assume that the result is true for $n = m$.

$$\text{i.e.,} \qquad \int_0^\infty e^{-st}(t^m f(t))\,dt = (-1)^m \frac{d^m}{ds^m}(F(s))$$

Now, it is necessary to prove that the result is true for $n = m + 1$.

For differentiating w.r.t. s

$$\frac{d}{ds}\int_0^\infty e^{-st}(t^m f(t))\,dt = (-1)^m \frac{d^{m+1}}{ds^{m+1}}F(s)$$

on by Leibnitz's rule

$$\int_0^\infty e^{-st}(-t)(t^m f(t))\,dt = (-1)^m \frac{d^{m+1}}{ds^{m+1}}F(s)$$

$$\therefore \qquad \int_0^\infty e^{-st}[t^{m+1}f(t)](dt = (-1)^{m+1}\frac{d^{m+1}}{ds^{m+1}}F(s)$$

Hence, the result is true for $n = m + 1$ also.

$\therefore$ by the principle of mathematical induction, the result is true for all positive integral values of n. Hence, the proof.

Cor 1: If $L\{f(t)\} = F(s)$ then

$$L\{t\,f(t)\} = -F'(s)$$

Cor 2: $L\{t\,f(t)\} = (-1)\dfrac{d}{ds}F(s)$

$$\therefore \qquad f(t) = \frac{1}{t}L^{-1}\left\{\frac{d}{ds}F(s)\right\}$$

$$= -\frac{1}{t}L^{-1}\{F'(s)\}$$

DIVISION BY *t*

If $L\{f(t)\} = F(s)$ then prove that

$$L\left\{\frac{f(t)}{t}\right\} = \int\limits_{s}^{\infty} F(s)\,ds$$

Proof: By definition

$$F(s) = \int\limits_{0}^{\infty} e^{-st} f(t)\,dt$$

Integrating both sides w.r.t. *s* from *s* to ∞

$$\therefore \qquad \int\limits_{s}^{\infty} F(s)\,ds = \int\limits_{s}^{\infty}\int\limits_{0}^{\infty} e^{-st} f(t)\,dt\,ds$$

Changing the order of double integration, as *s* and *t* are independent variables. We get

$$= \int\limits_{0}^{\infty}\left[\int\limits_{s}^{\infty} e^{-st}\,ds\right] f(t)\,dt$$

$$= \int\limits_{0}^{\infty}\left(\frac{e^{-st}}{-t}\right)_{s}^{\infty} f(t)\,dt$$

$$= \int\limits_{0}^{\infty}\left(0 - \frac{e^{-st}}{-t}\right) f(t)\,dt$$

$$= \int\limits_{0}^{\infty} e^{-st}\left(\frac{f(t)}{t}\right) dt$$

$$= L\left\{\frac{f(t)}{t}\right\}$$

$$\therefore \qquad L\left\{\frac{f(t)}{t}\right\} = \int\limits_{s}^{\infty} F(s)\,ds$$

Examples

Evaluate the following:

(1) $L\{t\cos at\}$

Solution:

Here $f(t) = \cos at$ $\qquad\qquad \therefore F(s) = \dfrac{s}{s^2 + a^2}$

$$\therefore \qquad L\{t\cos at\} = (-1)\frac{d}{ds}\frac{s}{s^2+a^2}$$

$$= -1\left[\frac{(s^2+a^2)1-s\times 2s}{(s^2+a^2)^2}\right]$$

$$= \frac{s^2-a^2}{(s^2+a^2)^2}$$

(2) $L\{t\sin at\}$

Solution:

Here $f(t)=\sin at \therefore F(s)=\dfrac{a}{s^2+a^2}$

$$\therefore \qquad L\{t\sin at\} = -1\frac{d}{ds}\left(\frac{a}{s^2+a^2}\right)$$

$$= -1(a)\frac{(-1)2s}{(s^2+a^2)^2} = \frac{2as}{(s^2+a^2)^2}$$

$$\therefore \qquad L\{t\sin at\} = \frac{2as}{(s^2+a^2)^2}$$

(3) $L\{t^3 e^{-3t}\}$

Solution:

Here $f(t)=e^{-3t} \quad \therefore F(s)=\dfrac{1}{s+3}$

$$\therefore \qquad L\{t^3 e^{-3t}\} = (-1)^3\frac{d^3}{ds^3}\left(\frac{1}{s+3}\right)$$

$$= \frac{(-1)(-1)(-2)(-3)}{(s+3)^4} = \frac{6}{(s+3)^4}$$

$$\therefore \qquad L\{t^3 e^{-3t}\} = \frac{6}{(s+3)^4}$$

(4) $L\{te^{-t}\cosh t\}$

Solution:

Here $f(t) = e^{-t}\cosh t$ $\therefore F(s) = \dfrac{s+1}{(s+1)^2 - 1^2} = \dfrac{s+1}{s^2 + 2s}$

$\therefore$

$$L\{te^{-t}\cosh t\} = (-1)\dfrac{d}{ds}\left(\dfrac{s+1}{s^2 + 2s}\right)$$

$$= -\left[\dfrac{(s^2 + 2s) - (s+1)(2s+2)}{(s^2 + 2s)^2}\right]$$

$$= \dfrac{s^2 + 2s + 2}{(s^2 + 2s)^2}$$

(5) Show that $L^{-1}\left\{\dfrac{1}{(s^2 + a^2)^2}\right\} = \dfrac{1}{2a^3}(\sin at - at\cos at)$

Solution:

We shall show that

$$L\{\sin at - at\cos at\} = \dfrac{2a^3}{(s^2 + a^2)^2}$$

Now, $L\{\sin at\} = \dfrac{a}{s^2 + a^2}$ and $L\{t\cos at\} = \dfrac{s^2 - a^2}{(s^2 + a^2)^2}$

$\therefore$

$$L\{\sin at - at\cos at\} = \dfrac{a}{s^2 + a^2} - \dfrac{a(s^2 - a^2)}{(s^2 + a^2)^2}$$

$$= \dfrac{a(s^2 + a^2) - a(s^2 - a^2)}{(s^2 + a^2)^2}$$

$$= \dfrac{2a^3}{(s^2 + a^2)^2}$$

(6) $L\left\{\dfrac{\sin at}{t}\right\}$

Solution:

Here $f(t) = \sin at$ $\therefore F(s) = \dfrac{a}{s^2 + a^2}$

$\therefore$

$$L\left\{\dfrac{\sin at}{t}\right\} = \int\limits_{s}^{\infty} F(s)\,ds$$

$$= \int_{s}^{\infty} \frac{a}{s^2 + a^2} \, ds = a \cdot \frac{1}{a} \tan^{-1} \frac{s}{a} \Bigg]_{s}^{\infty}$$

$$= \frac{\pi}{2} - \tan^{-1} \frac{s}{a} = \cot^{-1} \frac{s}{a}$$

(7) $L\left\{ \dfrac{e^{-at} - e^{-bt}}{t} \right\}$

Solution:

Here $f(t) = e^{-at} - e^{-bt}$ $\therefore F(s) = \dfrac{1}{s+a} - \dfrac{1}{s+b}$

$$\therefore \qquad L\left\{ \frac{e^{-at} - e^{-bt}}{t} \right\} = \int_{s}^{\infty} \left(\frac{1}{s+a} - \frac{1}{s+b} \right) ds$$

$$= [\log s + a - \log s + b]_{s}^{\infty}$$

$$= \left[\log \frac{s+a}{s+b} \right]_{s}^{\infty} = \log \frac{1 + \dfrac{a}{s}}{1 + \dfrac{b}{s}} \Bigg]_{s}^{\infty}$$

$$= \log 1 - \log \frac{1 + \dfrac{a}{s}}{1 + \dfrac{b}{s}}$$

$$= 0 - \log \frac{s+a}{s+b}$$

$$= \log \frac{s+b}{s+a}$$

(8) $L\left\{ \dfrac{\cos at - \cos bt}{t} \right\}$

Solution:

Here $f(t) = \cos at - \cos bt$ $\therefore F(s) = \dfrac{s}{s^2 + a^2} - \dfrac{s}{s^2 + b^2}$

$$\therefore \qquad L\left\{ \frac{\cos at - \cos bt}{t} \right\} = \int_{s}^{\infty} \left(\frac{s}{s^2 + a^2} - \frac{s}{s^2 + b^2} \right) ds$$

$$= \left[\frac{1}{2}\log(s^2 + a^2) - \frac{1}{2}\log(s^2 + b^2) \right]_s^\infty$$

$$= \frac{1}{2}\left[\log \frac{s^2 + a^2}{s^2 + b^2} \right]_s^\infty$$

$$= \frac{1}{2}\left[\log 1 - \log \frac{s^2 + a^2}{s^2 + b^2} \right]$$

$$= \frac{1}{2}\log \frac{s^2 + b^2}{s^2 + a^2}$$

(9) $L\left\{ \dfrac{1 - e^{-at}}{t} \right\}$

Solution:

Here
$$f(t) = 1 - e^{-at}, \quad \therefore F(s) = \frac{1}{s} - \frac{1}{s+a}$$

$$\therefore \qquad L\left\{ \frac{1 - e^{-at}}{t} \right\} = \int_s^\infty \left(\frac{1}{s} - \frac{1}{s+a} \right) ds$$

$$= \left[\log s - \log s + a \right]_s^\infty = \log \frac{s}{s+a} \bigg]_s^\infty$$

$$= \log 1 - \log \frac{s}{s+a}$$

$$= 0 + \log \frac{s+a}{s}$$

$$= \log \left(1 + \frac{a}{s} \right)$$

(10) $L\left\{ \dfrac{1 - \cos at}{t} \right\}$

Solution:

Here $f(t) = 1 - \cos at \quad \therefore F(s) = \dfrac{1}{s} - \dfrac{s}{s^2 + a^2}$

$$\therefore \quad L\left[\frac{1-\cos at}{t}\right] = \int_{s}^{\infty}\left(\frac{1}{s} - \frac{s}{s^2+a^2}\right)ds$$

$$= \left[\log s - \frac{1}{2}\log(s^2+a^2)\right]_{s}^{\infty}$$

$$= \log\frac{s}{\sqrt{s^2+a^2}}\Bigg]_{s}^{\infty}$$

$$= \log 1 - \log\frac{s}{\sqrt{s^2+a^2}}$$

$$= \log\frac{\sqrt{s^2+a^2}}{s} = \log\sqrt{1+\frac{a^2}{s^2}}$$

(11) $L\left\{\int_{0}^{t} e^{-x}\cos x\,dx\right\}$

Solution: We have

$$L\{e^{-x}\cos x\} = \frac{s+1}{(s+1)^2+1^2} = \frac{s+1}{s^2+2s+2}$$

But

$$L\left\{\int_{0}^{t} f(u)du\right\} = \frac{1}{s}F(s)$$

$$\therefore \quad L\left\{\int_{0}^{t} e^{-x}\cos x\,dx\right\} = \frac{1}{s}\cdot\frac{s+1}{s^2+2s+2} = \frac{s+1}{s^3+2s^2+2s}$$

(12) $L\left\{\int_{0}^{t} \frac{\sin x}{x}\,dx\right\}$

Solution: We have

$$L\left\{\frac{\sin x}{x}\right\} = \cot^{-1} s$$

$$\therefore \quad L\left\{\int_{0}^{t} \frac{\sin x}{x}\,dx\right\} = \frac{1}{s}\cot^{-1} s$$

(13) $L\left\{\int\limits_0^t x\cos x\,dx\right\}$

Solution: We have

$$L\{x\cos x\} = \frac{s^2-1}{(s^2+1)^2}$$

$$\therefore \qquad L\left\{\int\limits_0^t x\cos x\,dx\right\} = \frac{1}{s}\frac{s^2-1}{(s^2+1)^2}$$

(14) $L\left\{\int\limits_0^t e^{2x}x^3\,dx\right\}$

Solution: $f(t) = e^{2x}x^3 \quad \therefore F(s) = \dfrac{3!}{(s-2)^4}$

$$L\left\{\int\limits_0^t e^{2x}x^3\,dx\right\} = \frac{1}{s}\frac{3!}{(s-2)^4}$$

(15) $L\left\{\int\limits_0^t \frac{e^{-ax}-e^{-bx}}{x}\,dx\right\}$

Solution:

We have $L\left\{\dfrac{e^{-ax}-e^{-bx}}{x}\right\} = \log\dfrac{s+b}{s+a}$

$$\therefore \quad L\left\{\int\limits_0^t \frac{e^{-ax}-e^{-bx}}{x}\,dx\right\} = \frac{1}{s}\log\frac{s+b}{s+a}.$$

TRANSFORM OF A PERIODIC FUNCTION

A function $f(t)$ is said to be a periodic function of period $T > 0$ if

$$f(t) = f(t+T) = f(t+2T) =$$

or $\qquad\qquad f(t) = f(t+nT), \; n = 1, 2, 3, \ldots$

$\sin t$ and $\cos t$ are periodic functions of period 2π.

Theorem: Prove that the Laplace transform of a periodic function of period T is

$$L\{f(t)\} = \frac{1}{1 - e^{-sT}} \int_0^T e^{-st} f(t)\,dt$$

Proof: By definition

$$L\{f(t)\} = \int_0^\infty e^{-su} f(u)\,du$$

$$= \int_0^T e^{-su} f(u)\,du + \int_T^{2T} e^{-su} f(u)\,du + \int_{2T}^{3T} e^{-su} f(u)\,du + \ldots$$

$$= \sum_{n=0}^\infty \int_{nT}^{(n+1)T} e^{-su} f(u)\,du$$

Put $u = t + nT \therefore du = dt$

When $u = nT,\ t = 0$

$u = (n+1)T,\ t = T$

$$= \sum_{n=0}^\infty \int_0^T e^{-s(t+nT)} f(t+nT)\,dt$$

$$= \sum_{n=0}^\infty (e^{-sT})^n \int_0^T e^{-st} f(t)\,dt \qquad\qquad \because f(t+nT) = f(t)$$

$$= (1 + e^{-sT} + e^{-2T} + e^{-3sT} \ldots) \int_0^T e^{-st} f(t)\,dt$$

$$= (1 - e^{-sT})^{-1} \int_0^T e^{-st} f(t)\,dt \qquad\qquad (\because \text{ by binomial theorem})$$

$$= \frac{1}{1 - e^{-sT}} \int_0^T e^{-st} f(t)\,dt$$

$$\therefore \qquad L\{f(t)\} = \frac{1}{1 - e^{-sT}} \int_0^T e^{-st} f(t)\,dt$$

Examples

(1) If $f(t) = t^2 \quad 0 < t < 2$

$$= 0 \text{ for } t > 2$$

and $f(t + 2) = f(t)$, find $L\{f(t)\}$.

Solution: Here $T = 2$

$$L\{f(t)\} = \frac{1}{1 - e^{-2s}} \int_0^2 e^{-st} t^2 \, dt$$

$$= \frac{1}{1 - e^{-2s}} \left[t^2 \frac{e^{-st}}{-s} - 2t \frac{e^{-st}}{s^2} + 2 \frac{e^{-st}}{-s^3} \right]_0^2$$

$$= \frac{1}{1 - e^{-2s}} \left[\frac{4e^{-2s}}{-s} - \frac{4e^{-2s}}{s^2} + \frac{2e^{-2s}}{-s^3} - \frac{2}{-s^3} \right]$$

$$= \frac{2}{s^3(1 - e^{-2s})} \left[1 - (1 + 2s + 2s^2)e^{-2s} \right]$$

(2) A periodic function of period $\dfrac{2\pi}{w}$ is defined by

$$f(t) = E \sin wt \qquad 0 < t < \frac{\pi}{w}$$

$$= 0 \qquad \frac{\pi}{w} < t \le \frac{2\pi}{w}$$

where E and w are constants. Show that

$$L\{f(t) = \frac{Ew}{(s^2 + w^2)(1 - e^{-\frac{\pi s}{w}})}$$

Solution: The given function is periodic with period $\dfrac{2\pi}{w}$.

$$\therefore \qquad L\{f(t)\} = \frac{1}{1 - e^{-\frac{2\pi s}{w}}} \int_0^{2\pi/w} e^{-st} f(t) \, dt$$

$$= \frac{1}{1-e^{-\frac{2\pi s}{w}}} \int_0^{\pi/w} e^{-st} E \sin wt \, dt + 0$$

$$= \frac{E}{1-e^{-2\pi s/w}} \left[e^{-st} \frac{(-s\sin wt - w\cos wt)}{s^2+w^2} \right]_0^{\pi/w}$$

$$= \frac{Ew}{1-e^{-\frac{2\pi s}{w}}} \frac{(1+e^{-\frac{\pi s}{w}})}{s^2+w^2}$$

$$= \frac{Ew(1+e^{-\frac{\pi s}{w}})}{(1-e^{-\frac{\pi s}{w}})(1+e^{-\frac{\pi s}{w}})(s^2+w^2)}$$

$$\therefore \quad L\{f(t)\} = \frac{Ew}{(1-e^{-\frac{\pi s}{w}})(s^2+w^2)}$$

(3) A periodic function $f(t)$ of period $2a$, $a > 0$ is defined by

$$f(t) = a \quad \text{for} \quad 0 \le t \le a$$
$$= -a \quad \text{for} \quad a < t \le 2a$$

Show that $L\{f(t)\} = \dfrac{a}{s}\tanh\left(\dfrac{as}{2}\right)$.

Solution: Since the given funtion $f(t)$ is periodic with period $2a$, we have

$$L\{f(t)\} = \frac{1}{1-e^{-2as}} \int_0^{2a} e^{-st} f(t) \, dt$$

$$= \frac{1}{1-e^{-2as}} \left[\int_0^a a e^{-st} \, dt + \int_a^{2a} (0a)e^{-st} \, dt \, 0 \right]$$

$$= \frac{a}{1-e^{-2as}} \left[\left(\frac{e^{-st}}{-s}\right)_0^a - \left(\frac{e^{-st}}{-s}\right)_a^{2a} \right]$$

$$= \frac{a}{s(1-e^{-2as})} (1 - e^{-as} - e^{-as} + e^{-2as}$$

$$= \frac{a(1-e^{-as})^2}{s(1-e^{-as})(1+e^{-as})} = \frac{a}{s}\frac{1-e^{-as}}{1+e^{-as}}$$

$$= \frac{a}{s}\frac{e^{\frac{as}{2}}-e^{-\frac{as}{2}}}{e^{\frac{as}{2}}+e^{\frac{-as}{2}}} = \frac{a}{s}\tanh\left(\frac{as}{2}\right)$$

(4) Find the Laplace transform of the triangular wave function of period $2a$ given by

$$f(t) = t \qquad 0 < t < a$$
$$= 2a - t \quad a < t < 2a$$

Solution: Given function $f(t)$ is a periodic function of period $T = 2a$.

$$\therefore \quad L\{f(t)\} = \frac{1}{1-e^{-2as}}\int_0^{2a} e^{-st} f(t)dt$$

$$= \frac{1}{1-e^{-2as}}\left[\int_0^a te^{-st}dt + \int_a^{2a}(2a-t)e^{-st}dt\right]$$

$$= \frac{1}{1-e^{-2as}}\left[\left(t\frac{e^{-st}}{-s}-\frac{e^{-st}}{s^2}\right)_0^a + \left((2a-t)\frac{e^{-st}}{-s}-(-1)\frac{e^{-st}}{s^2}\right)_a^{2a}\right]$$

$$= \frac{1}{1-e^{-2as}}\left[\frac{ae^{-as}}{-s}-\frac{e^{-as}}{s^2}+\frac{1}{s^2}+\frac{e^{-2as}}{s^2}-\frac{ae^{-as}}{-s}-\frac{e^{-as}}{s^2}\right]$$

$$= \frac{1}{s^2(1-e^{-2as})}(1-2e^{-2as}+e^{-2as})$$

$$= \frac{1}{s^2}\frac{(1-e^{-as})^2}{(1-e^{-as})(1+e^{-as})}$$

$$= \frac{1}{s^2}\frac{(1-e^{-as})}{(1+e^{-as})} = \frac{1}{s^2}\frac{e^{\frac{as}{2}}-e^{-\frac{as}{2}}}{e^{\frac{as}{2}}+e^{-\frac{as}{2}}}$$

$$= \frac{1}{s^2}\tanh\left(\frac{as}{2}\right)$$

(5) Find the Laplace transform of a periodic function with period a is given by

$$f(t) = E \qquad 0 < t < \frac{a}{2}$$

$$= -E \qquad \frac{a}{2} < t < a$$

Solution: Here $T = a$ and

$$L\{f(t)\} = \frac{1}{1 - e^{-as}} \int_0^a e^{-st} f(t) dt$$

$$= \frac{1}{1 - e^{-as}} \left[\int_0^{a/2} E e^{-st} dt + \int_{a/2}^a (-E) e^{-st} dt \right]$$

$$= \frac{E}{1 - e^{-as}} \left[\left(\frac{e^{-st}}{-s} \right)_0^{a/2} - \left(\frac{e^{-st}}{-s} \right)_{a/2}^a \right]$$

$$= \frac{E}{s} \frac{(1 - 2e^{-\frac{as}{2}} + e^{-as})}{(1 - e^{-as})} = \frac{E}{s} \frac{(1 - e^{-\frac{as}{2}})^2}{(1 - e^{-\frac{as}{2}})(1 + e^{-\frac{as}{2}})}$$

$$= \frac{E}{s} \frac{(1 - e^{-\frac{as}{2}})}{(1 + e^{-\frac{as}{2}})} = \frac{E}{s} \frac{e^{\frac{as}{4}} - e^{-\frac{as}{4}}}{e^{\frac{as}{4}} + e^{-\frac{as}{4}}}$$

$$= \frac{E}{s} \tanh\left(\frac{as}{4} \right)$$

Transform of the Step Function

In many applications, we deal with the discontinuous function $H(t - a)$, defined as follows:

$$H(t - a) = 0 \text{ for } t \le a$$

$$= 1 \text{ for } t > a \tag{1}$$

where a is a non-negative constant.

This function is known as the unit *step function* or the *Heaviside unit function*. The graph of this function is as follows:

From the graph of $H(t - a)$, we note that the value of this function suddenly jumps (steps up) from the value zero to

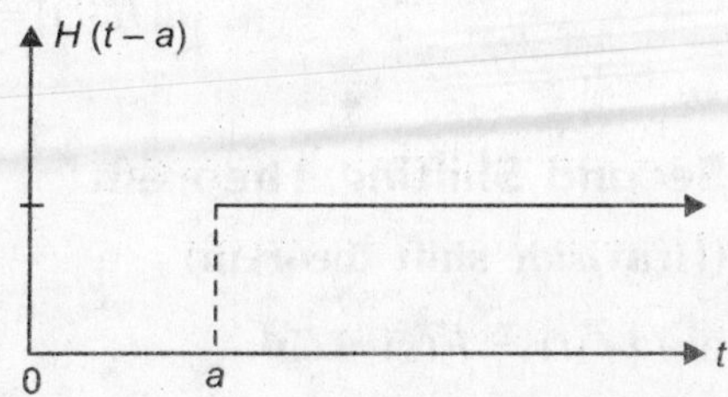

the value 1 as $t > a$ from the left, and retains the value 1 for all $t > a$. This is why, $H(t-a)$ is called the unit step function. In the particular case when $a = 0$ the function $H(t-a)$ becomes $H(t)$

where $$H(t)= 0 \text{ for } t < 0$$
$$= 1 \text{ for } t > 0$$

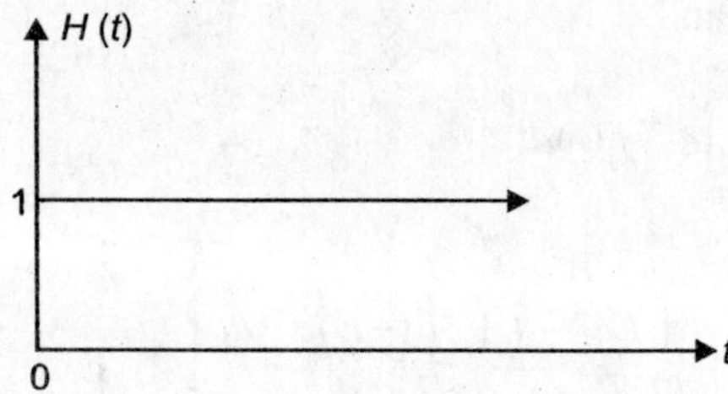

Laplace Transform of the Function $H(t-a)$

By definition of Laplace transform

$$L\{H(t-a)\} = \int_0^\infty e^{-st} H(t-a)dt$$

$$= \int_0^a e^{-st}(0)dt + \int_a^\infty e^{-st}(1)dt$$

$$= \int_a^\infty e^{-st}dt$$

$$= \left[\frac{e^{-st}}{-s}\right]_a^\infty = \frac{1}{s}e^{-as}$$

$$\therefore \quad L\{H(t-a)\} = \frac{1}{s}e^{-as}$$

Cor: For $a = 0$

$$L\{H(t)\} = \frac{1}{s}e^{0s} = \frac{1}{s}$$

Second Shifting Theorem

(Heaviside shift theorem)

If $L\{f(t)\} = F(s)$ then

$$L\{f(t-a)H(t-a)\} = e^{-as}L\{f(t)\}$$

$$= e^{-as}F(s)$$

Proof: By the definition of Laplace transform

$$L\{f(t-a)H(t-a)\} = \int_0^\infty e^{-st}f(t-a)H(t-a)dt$$

$$= \int_0^a e^{-st}f(t-a)(0)dt + \int_a^\infty e^{-st}f(t-a)(1)dt$$

$$= \int_a^\infty e^{-st}f(t-a)dt$$

Put
$$u = t-a \quad \bigg| \quad t = a \qquad u = 0$$
$$du = dt \quad \bigg| \quad t = \infty \qquad u = \infty$$

$$= \int_0^\infty e^{-s(u+a)}f(u)du$$

$$= e^{-as}\int_0^\infty e^{-su}f(u)du$$

$$= e^{-as}F(s)$$

$$\therefore \qquad L\{f(t-a)H(t-a)\} = e^{-as}F(s)$$

Cor 1: Put $a = 0$

$$L\{f(t)H(t)\} = F(s)$$

Cor 2: If $f(t) = 1$ then $L\{H(t-a)\} = e^{-as}L\{1\}$

$$= \frac{1}{s}e^{-as}$$

Theroem: If a function $f(t)$ is defined by

$$f(t) = f_1(t) \quad \text{for} \quad t \le a$$
$$= f_2(t) \quad \text{for} \quad t > a$$

Then

$$f(t) = f_1(t) + \{f_2(t) - f_1(t)\}H(t-a)$$

Proof: By the definition of $H(t-a)$, we have

$$\{f_2(t) - f_1(t)\}H(t-a) = \begin{vmatrix} f_2(t) - f_1(t) & t > a \\ 0 & t \le a \end{vmatrix}$$

Adding $f_1(t)$ to both sides, this becomes

$$f_1(t) + \{f_2(t) - f_1(t)\}H(t-a) = \begin{vmatrix} f_1(t) + \{f_2(t) - f_1(t)\} & \text{for } t > a \\ f_1(t) + 0 & \text{for } t \le a \end{vmatrix}$$

$$= \begin{vmatrix} f_2(t) & \text{for } t > a \\ f_1(t) & \text{for } t \le a \end{vmatrix}$$

$$= f(t) \qquad \text{by definition}$$

$$\therefore \qquad f(t) = f_1(t) + \{f_2(t) - f_1(t)\}H(t-a)$$

TRANSFORMS OF UNIT IMPULSE FUNCTION

When a very large force acts for a very short time, then such forces are called impulsive forces. In mechanics, we come across problems where a very large force acts for a very short time. In the study of bending of beams, we have point loads which is equivalent to large pressure acting over a very small area. To deal with such problems, we introduce the unit impulse function or Dirac-delta function.

Definition: This unit impulse function is defined as the limiting form of the following function as $\in \to 0$.

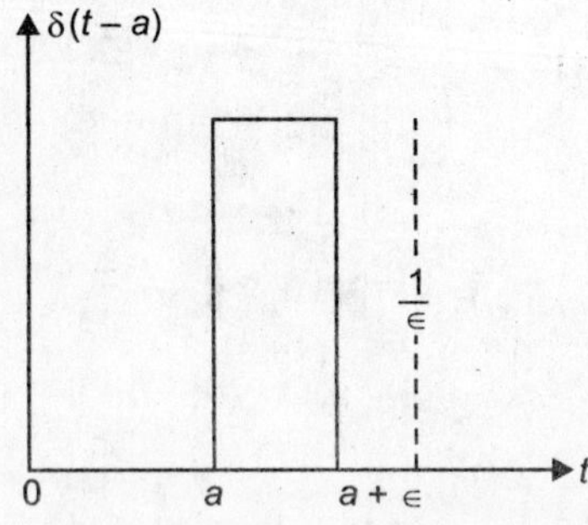

$$\delta_\in(t-a) = \begin{vmatrix} 0 & \text{if } t < a \\ \dfrac{1}{\in} & \text{if } a < t < a+\in \\ 0 & \text{if } t > a+\in \end{vmatrix}$$

The graph of this function is as shown above. As $\in \to 0$, the height of the rectangular region increases and width decreases in such a way that the area is always equal to 1.

Thus, the unit impulse function is

$$\delta(t-a) = \begin{vmatrix} 0 & \text{if } t < a \\ \infty & \text{if } t = a \\ 0 & \text{if } t > a \end{vmatrix}$$

such that
$$\int_0^\infty \delta(t-a)\,dt = 1 \, .$$

Laplace transform of unit impulse function $\delta(t-a)$

Consider

$$L\{\delta_\in(t-a)\} = \int_0^\infty e^{-st}\delta_\in(t-a)\,dt$$

$$= \int_a^a e^{-st}(0)\,dt + \int_a^{a+\in} e^{-st}\left(\frac{1}{\in}\right)dt + \int_{a+\in}^\infty e^{-st}(0)\,dt$$

$$= \int_a^{a+\in} e^{-st}\frac{1}{\in}\,dt$$

$$= \left[\frac{1}{\in}\frac{e^{-st}}{-s}\right]_a^{a+\in} = -\frac{1}{\in s}[e^{-s(a+\in)} - e^{-as}]$$

$$= e^{-as}\left[\frac{1-e^{-\in s}}{\in s}\right]$$

By L' Hospital's rule $\lim\dfrac{1-e^{-\in s}}{\in s} = 1$

Hence, by taking limit on both sides as $\in \to 0$, we get

$$L\{\delta(t-a)\} = e^{-as}$$

Examples

(1) Evaluate $L\{t^2 H(t-3)\}$

Solution:

We can write t^2 as

$$t^2 = [(t-3)+3]^2$$

$$= (t-3)^2 + 6(t-3) + 9$$

$$\therefore \qquad L\{t^2 H(t-3)\}$$

$$= L\{(t-3)^2 + 6(t-3) + 9\}H(t-3)$$

$$= f(t-3)H(t-3)$$

$$\therefore \quad L\{t^2 H(t-3)\} = e^{-3s} L\{t^2 + 6t + 9\}$$

$$= e^{-3s}\left(\frac{2}{s^3} + \frac{6}{s^2} + \frac{9}{s}\right)$$

(2) Evaluate $L\{e^{(t-2)} H(t-2)\}$

Solution:

$$L\{e^{(t-2)} H(t-2)\} = e^{-2s} L\{e^t\}$$

$$= e^{-2s}\frac{1}{s-1}$$

(3) Express the following function in terms of the unit step function and find $L\{f(t)\}$.

$$f(t) = 2t \quad 0 < t < \pi$$
$$= 1 \quad t > \pi$$

Solution: By theorem,

$$f(t) = f_1(t) + [f_2(t) - f_1(t)]H(t-\pi)$$

$$= 2t + [1 - 2t]H(t-\pi)$$

$$= 2t + [1 - 2(t-\pi) - 2\pi]H(t-\pi)$$

$$= 2t + [(1 - 2\pi) - 2(t-\pi)]H(t-\pi)$$

$$\therefore \quad L\{f(t)\} = L\{2t\} + e^{-\pi s} L(1 - 2\pi)\} - e^{-\pi s} L\{2t\}$$

$$= 2\frac{1}{s^2} + (1 - 2\pi)e^{-\pi s}\frac{1}{s} - 2e^{-\pi s}\cdot\frac{1}{s^2}$$

$$= \frac{2}{s^2} + (2 - 2\pi)\frac{e^{-\pi s}}{s} - \frac{2e^{-\pi s}}{s^2}$$

(4) Express the following function in terms of Heaviside's unit step function and find its L.T.

$$f(t) = \begin{vmatrix} t^2 & 0 < t < 2 \\ 4t & t > 2 \end{vmatrix}$$

Solution: Given function $f(t)$ can be expressed in terms of the unit step function as

$$f(t) = f_1(t) + \{f_2(t) - f_1(t)\}H(t-2)$$

$$= t^2 + \{4t - t^2\}H(t-2)$$

$$= t^2 + \{4(t-2+2) - (t-2+2)^2\}H(t-2)$$

$$= t^2 + \{4(t-2) + 8 - (t-2)^2 - 4(t-2) - 4\}H(t-2)$$

$$= t^2\{4 - (t-2)^2\}H(t-2)$$

$$\therefore \qquad L\{f(t) = L\{t^2\} + e^{-2s}L\{4 - t^2\}$$

$$= \frac{2}{s^3} + e^{-2s}\left(\frac{4}{s} - \frac{2}{s^3}\right)$$

(5) Express the following function in terms of unit step function and find its L.T.

$$f(t) = \begin{vmatrix} t^2 & 0 < t < 2 \\ 4t & 2 < t < 4 \\ 8 & t > 4 \end{vmatrix}$$

Solution: The given function can be written as

$$f(t) = t^2 + \{4t - t^2\}H(t-2) + \{8 - 4t\}H(t-4)$$

$$= t^2 + \{4 - (t-2)^2\}H(t-2) + \{8 - 4(t-4+4)\}H(t-4)$$

$$= t^2 + \{4 - (t-2)^2\}H(t-2) + \{-4(t-4) - 8\}H(t-4)$$

$$\therefore \qquad L\{f(t)\} = L\{t^2\} + e^{-2s}L\{4 - t^2\} - e^{-4s}L\{4t + 8\}$$

$$= \frac{2}{s^3} + e^{-2s}\left(\frac{4}{s} - \frac{2}{s^3}\right) - 4e^{-4s}\left(\frac{1}{s^2} + \frac{2}{s}\right)$$

(6) Express $f(t)$ in terms of Heaviside's unit step function and find its L.T.

$$f(t) = \begin{vmatrix} \sin t & 0 < t < \pi \\ \sin 2t & \pi < t < 2\pi \\ \sin 3t & t > 2\pi \end{vmatrix}$$

Solution: The given function can be written as

$$f(t) = f_1(t) + \{f_2(t) - f_1(t)\}1 + (t - \pi) + (f_3(t) - f_2(t))\}H(t - 2\pi)$$

$$= \sin t + \{\sin 2t - \sin t\}H(t-\pi) + \{\sin 3t - \sin 2t\}H(t-2\pi)$$

$$= \sin t + [\sin 2(t-\pi) - \sin(t-\pi)]H(t-\pi)$$

$$+[\sin 3(t-2\pi) - \sin 2(t-2\pi)]H(t-2\pi)$$

$$\therefore \quad L\{f(t)\} = L\{\sin t\} + e^{-\pi s}L\{\sin 2t - \sin t\} + e^{-2\pi s}L\{\sin 3t - \sin 2t\}$$

$$= \frac{1}{s^2+1} + e^{-\pi s}\left(\frac{2}{s^2+2^2} - \frac{1}{s^2+1}\right) + e^{-2\pi s}\left(\frac{3}{s^2+3^2} - \frac{2}{s^2+2^2}\right)$$

$$= \frac{1}{s^2+1} + e^{-\pi s}\left(\frac{2}{s^2+4} - \frac{1}{s^2+1}\right) + e^{-2\pi s}\left(\frac{3}{s^2+9} - \frac{2}{s^2+4}\right).$$

(7) Express the following function in terms of unit step function and find its L.T.

$$f(t) = \begin{vmatrix} \cos t & \text{for } 0 < t < \pi \\ \sin t & \text{for } t > \pi \end{vmatrix}$$

Solution: Given function can be expressed as

$$f(t) = f_1(t) + \{f_2(t) - f_1(t)\}H(t-\pi)$$

$$= \cos t + \{\sin t - \cos t\}H(t-\pi)$$

$$= \cos t + \{(\sin t \cos \pi - \cos t \sin \pi)$$

$$+(\cos t \cos \pi + \sin t \sin \pi)\}H(t-\pi)$$

$$= \cos t + \{-\sin(t-\pi) + \cos(t-\pi)\}H(t-\pi)$$

$$\therefore \qquad L\{f(t)\} = L\{\cos t\} + e^{-\pi s}L\{-\sin t + \cos t\}$$

$$= \frac{s}{s^2+1} + e^{-\pi s}\left(\frac{-1}{s^2+1} + \frac{s}{s^2+1}\right)$$

(8) Express the function in terms of Heaviside's unit step function and find its L.T.

$$f(t) = \begin{vmatrix} \sin t & \text{for } 0 \le t \le \pi \\ t & \text{for } \pi \le t \le 2\pi \\ \cos t & \text{for } t > 2\pi \end{vmatrix}$$

Solution: Given function can be expressed as

$$f(t) = f_1(t) + \{f_2(t) - f_1(t)\}H(t-\pi) + \{f_3(t) - f_2(t)\}H(f-2\pi)$$

$$= \sin t + \{t - \sin t\}H(t-\pi) + \{\cos t - t\}H(t-2\pi)$$

$$= \sin t + \{\pi + (t - \pi) + \sin(t - \pi)\}H(t - \pi)$$

$$+\{-2\pi - (t - 2\pi) + \cos(t - 2\pi)\}H(t - 2\pi)$$

$$\therefore \quad L\{f(t)\} = L\{\sin t\} + e^{-\pi s}L\{\pi + t + \sin t\} + e^{-2\pi s}L\{\cos t - t - 2\pi\}$$

$$= \frac{1}{s^2 + 1} + e^{-\pi s}\left(\frac{\pi}{s} + \frac{1}{s^2} + \frac{1}{s^2 + 1}\right) + e^{-2\pi s}\left(\frac{s}{s^2 + 1} - \frac{1}{s^2} - \frac{2\pi}{s}\right)$$

(9) Evaluate $\displaystyle\int_0^\infty t e^{-2t} \sin t\, dt$

Solution: We have

$$\int_0^\infty e^{-2t}(t \sin t)dt = \int_0^\infty e^{-st}(t \sin t)dt$$

where $s = 2$

$$= L\{t \sin t\} \text{ by definition}$$

$$= (-1)\frac{d}{ds}\left(\frac{1}{s^2 + 1}\right)$$

$$= \frac{2s}{(s^2 + 1)^2} = \frac{2 \times 2}{(2^2 + 1)^2} = \frac{4}{25}.$$

(10) Evaluate $\displaystyle L\left\{\int_0^t \frac{e^t \sin t}{t} dt\right\}$

Solution: We have

$$L\left\{\frac{\sin t}{t}\right\} = \int_0^\infty \frac{ds}{s^2 + 1} = [\tan^{-1} s]_0^\infty = \frac{\pi}{2} - \tan^{-1} s$$

$$= \cot^{-1} s$$

$$\therefore \qquad L\left\{e^t\left(\frac{\sin t}{t}\right)\right\} = \cot^{-1}(s - 1) \qquad \text{by shifting property}$$

$$\therefore \qquad L\left[\int_0^t\left\{e^t\left(\frac{\sin t}{t}\right)dt\right\}\right] = \frac{1}{s}\cot^{-1}(s - 1)$$

QUESTION BANK 5

(A) Questions carrying one mark

Choose the correct answers from the given alternatives.

(1) Laplace transform of $f(t), t \geq 0$ is defined by

(a) $\int\limits_{0}^{\infty} e^{-st} f(t)\,dt$

(b) $\int\limits_{0}^{\infty} e^{st} f(t)\,dt$

(c) $\int\limits_{-\infty}^{\infty} e^{-st} f(t)\,dt$

(d) $d = \int\limits_{1}^{n} e^{-st} f(t)\,dt$

(2) Laplace transform of a^t is

(a) $\dfrac{1}{s}$

(b) $\dfrac{1}{s - \log a}$

(c) $\dfrac{1}{\log a}$

(d) $e^{-s} \dfrac{1}{a}$

(3) Laplace transform of $\cos 2t$ is

(a) $\dfrac{1}{s^2 + 2^2}$

(b) $\dfrac{s}{s^2 + 2^2}$

(c) $\dfrac{2}{s^2 + 2^2}$

(d) $\dfrac{s^2}{s^2 + 2^2}$

(4) Laplace transform of $f'(t)$ is

(a) $SF(s) - f(0)$

(b) $SF'(s) - f(0)$

(c) $f(s) - f(0)$

(d) $\bar{F}(s)$

(5) Laplace transform of $e^{2t}\, t^3$ is

(a) $\dfrac{3}{(s-3)^3}$

(b) $\dfrac{s}{(s-2)^3}$

(c) $\dfrac{6}{(s-2)^4}$

(d) $\dfrac{1}{s^4}$

(6) Laplace transform of $L\left\{ \int\limits_{0}^{t} f(x)\,dx \right\}$ is

(a) $SF(s)$

(b) $\dfrac{1}{s} F(s)$

(c) $F'(s)$

(d) $\dfrac{F'(s)}{s}$

(7) Laplace transform of $t \cos at$ is

(a) $\dfrac{s^2 - a^2}{(s^2 + a^2)^2}$

(b) $\dfrac{s^2 + a^2}{(s^2 - a^2)^2}$

(c) $\dfrac{(s^2 + a^2)^2}{s^2 - a^2}$

(d) $\dfrac{s}{s^2 + a^2}$

(8) Laplace transform of $t \sin 2t$ is

(a) $\dfrac{2}{(s^2 + a^2)^2}$

(b) $\dfrac{4s}{(s^2 + 4)^2}$

(c) $\dfrac{s}{(s^2 + 4)^2}$

(d) $\dfrac{s}{(s^2 - a^2)^2}$

(9) Laplace transform of $\dfrac{f(t)}{t}$ is

(a) $\displaystyle\int_s^\infty F(s)\,ds$

(b) $\displaystyle\int_0^\infty F(s)\,ds$

(c) $\displaystyle\int_1^\infty \dfrac{1}{s} F(s)\,ds$

(d) $\displaystyle\int_{-\infty}^\infty \dfrac{F(s)}{s}\,ds$

(10) Laplace transform of $\displaystyle\int_0^t e^{-t} \cos t\, dt$ is

(a) $\dfrac{1}{s} \cdot \dfrac{s+1}{s^2 + 2s + 2}$

(b) $\dfrac{s+1}{s^2 + 2s + 2}$

(c) $\dfrac{s}{s^2 + 2s + 2}$

(d) $\dfrac{s^2 + 1}{s^2 + 2s + 1}$

(11) Laplace transform of $\displaystyle\int_0^t \dfrac{\sin t}{t}\,dt$ is

(a) $\cot^{-1} s$

(b) $\dfrac{1}{s}\cot^{-1} s$

(c) $\dfrac{1}{s^2 + 1}$

(d) $\dfrac{s}{s^2 + 1}$

(12) Laplace transform of unit step function $H(t-a)$ is

    (a) $\dfrac{1}{s}F(s)$ (b) $\dfrac{1}{s}e^{-as}$

    (c) $\dfrac{1}{s}$ (d) e^{-as}

(13) By Heaviside shift theorem Laplace transform of $f(t-a)H(t-a)$ *is*

    (a) $e^{-as}F(s)$

    (b) $\dfrac{1}{s}F(s)$

    (c) $\dfrac{e^{-as}}{s}$

    (d) $e^{as}F(s)$

(14) Laplace transform unit impulse function $\delta(t-a)$ is

    (a) $\dfrac{1}{s}$ (b) e^{-as}

    (c) $\dfrac{1}{s}e^{-as}$ (d) $\dfrac{1}{s}e^{as}$

(15) Laplace transform of $e^{(t-1)}H(t-1)$ is

    (a) $\dfrac{e^{-s}}{s-1}$ (b) $\dfrac{e^{s}}{s+1}$

    (c) $\dfrac{1}{s+1}$ (d) e^{-s}

(16) Laplace transform of unit step function $tH(t-3)$ is

    (a) $\dfrac{e^{-3}s}{s-1}$ (b) $e^{-3s}\dfrac{1}{s^2}$

    (c) $e^{-3s}\dfrac{1}{s^3}$ (d) se^{-3s}

Answers for A

(1) a	(2) b	(3) b	(4) a	(5) c
(6) b	(7) a	(8) b	(9) a	(10) a
(11) b	(12) b	(13) a	(14) b	(15) a
(16) b				

(B) (1) Find the Laplace transform of

    (i) $f(t) = e^{at}$

    (ii) $f(t) = a^t$

(2) Prove that

    (i) $L\{\sin at\} = \dfrac{a}{s^2 + a^2}$

    (ii) $L\{\cos at\} = \dfrac{s}{s^2 + a^2}$

(3) Prove that $L\{t^n\} = \dfrac{n!}{s^{n+1}}$ where n is a positive integer.

(4) If $f(t) = \begin{vmatrix} 2t & \text{for} & 0 \le t \le 5 \\ 1 & \text{for} & t > 5 \end{vmatrix}$ Find $L\{f(t)\}$

(5) If $L\{f(t)\} = F(s)$ then show that

$$L\left\{ \int_0^t f(u)\,du \right\} = \frac{1}{s} F(s)$$

(6) Find the Laplace transform of
    (i) $\sin 2t \sin 3t$
    (ii) $\cos 3t \cos 4t$
    (iii) $\sin 4t \cos 2t.$

(7) Find $L\{e^{-t}(3\sinh 2t - 2\cosh 3t)\}$

(8) Find (i) $L\{\sin^3 2t\}$ (ii) $L\{\cos^3 2t\}$

(9) Find (i) $L\{e^{-3t}\sin 2t\}$ (ii) $L\{e^{-3t}\cos 2t\}$

(10) Find (i) $L\{\cos^2 4t\}$ (ii) $L\{te^{-t}\sin t\}$

(11) Find (i) $L\{te^{-t}\cos 2t\}$ (ii) $L\{t^3 e^{-2t}\}$

(12) Find (i) $L\{e^{-t}\sin^2 t\}$ (ii) $L\{e^{2t}\cos 2t\}$

(13) Find (i) $L\{t\sin at\}$ (ii) $L\{t\cos at\}$

(14) Find (i) $L\left\{\dfrac{1-e^{at}}{t}\right\}$ (ii) $L\left\{\dfrac{1-\cos at}{t}\right\}$

(15) Find $L\left\{\dfrac{\cos at - \cos bt}{t}\right\}$

(16) Find $L\{e^{-2t}(3\cos t - 2\sin 5t)\}$

(17) Find the Laplace transform of $f(t) = \begin{cases} t/\lambda & \text{when} \quad 0 < t < \lambda \\ 0 & \text{when} \quad t > \lambda \end{cases}$

(18) If $L\{f(t)\} = F(s)$ then prove that

$$L\{t^n f(t)\} = (-1)^n \frac{d^n}{ds^n}\{F(s)\} \qquad n = 1, 2, 3, \ldots$$

(19) Show that   (i) $L\{e^{at} \sin bt\} = \dfrac{b}{(s-a)^2 + b^2}$

$$\text{(ii) } L\{e^{at} \cos bt\} = \frac{s-a}{(s-a)^2 + b^2}$$

(20) Find $L\left\{ \dfrac{e^{-at} - e^{-bt}}{t} \right\}$

(21) Find $L\left\{ \displaystyle\int_0^t e^{-x} \cos x \, dx \right\}$

(22) $L\left\{ \displaystyle\int_0^t \dfrac{\sin x}{x} \, dx \right\}$

(23) Find $L\left\{ \displaystyle\int_0^t x \cos x \, dx \right\}$

(24) Find $L\left\{ \displaystyle\int_0^t e^{2x} x^3 \, dx \right\}$

UNIT 6

Laplace Transforms-2

INVERSE LAPLACE TRANSFORMS

Suppose $L\{f(t)\} = F(s)$ then $f(t)$ is called the inverse Laplace transform of $F(s)$ and is written as $L^{-1}\{F(s)\} = f(t)$ Here L^{-1} denotes the inverse Laplace transform. The inverse Laplace transforms given below follow at once from the results of Laplace transforms studied earlier.

$$(1) \quad L^{-1}\left\{\frac{1}{s}\right\} = 1$$

$$(2) \quad L^{-1}\left\{\frac{1}{s-a}\right\} = e^{at}$$

$$(3) \quad L^{-1}\left\{\frac{1}{s+a}\right\} = e^{-at}$$

$$(4) \quad L^{-1}\left\{\frac{a}{s^2+a^2}\right\} = \sin at$$

$$(5) \quad L^{-1}\left\{\frac{s}{s^2+a^2}\right\} = \cos at$$

$$(6) \quad L^{-1}\left\{\frac{2as}{(s^2+a^2)^2}\right\} = t\sin at$$

$$(7) \quad L^{-1}\left\{\frac{2a^3}{(s^2+a^2)^2}\right\} = \sin at - at\cos at$$

$$(8) \quad L^{-1}\left\{\frac{a}{s^2-a^2}\right\} = \sinh at$$

(9) $\quad L^{-1}\left\{\dfrac{s}{s^2-a^2}\right\}=\cosh at$

(10) $\quad L^{-1}\left\{\dfrac{n!}{s^{n+1}}\right\}=t^n \quad n=1,\,2,\,3,\,...$

(11) $\quad L^{-1}\left\{\dfrac{n!}{(s-a)^{n+1}}\right\}=e^{at}t^n$

(12) $\quad L^{-1}\left\{\dfrac{b}{(s-a)^2+b^2}\right\}=e^{at}\sin bt$

(13) $\quad L^{-1}\left\{\dfrac{s-a}{(s-a)^2+b^2}\right\}=e^{at}\cos bt$

(14) $\quad L^{-1}\left\{\dfrac{b}{(s-a)^2-b^2}\right\}=e^{at}\sinh bt$

(15) $\quad L^{-1}\left\{\dfrac{s-a}{(s-a)^2-b^2}\right\}=e^{at}\cosh bt$

Partial fractions

(1) $\quad \dfrac{1}{(s+a)(s+b)}=\dfrac{A}{s+a}+\dfrac{B}{s+b}$

(2) $\quad \dfrac{1}{(s+a)(s+b)^2}=\dfrac{A}{s+a}+\dfrac{B}{(s+b)}+\dfrac{C}{(s+b)^2}$

(3) $\quad \dfrac{1}{(s+a)(s^2+bs+c)}=\dfrac{A}{s+a}+\dfrac{Bs+C}{s^2+bs+c}$

Examples

Find the inverse Laplace transform of

(1) $\quad L^{-1}\left\{\dfrac{3s+5\sqrt{2}}{s^2+8}\right\}$

Solution: We have

$$L^{-1}\left\{\dfrac{3s+5\sqrt{2}}{s^2+8}\right\}=3L^{-1}\left\{\dfrac{s}{s^2+(2\sqrt{2})^2}\right\}+5\sqrt{2}\,L^{-1}\left\{\dfrac{1}{s^2+(2\sqrt{2})^2}\right\}$$

$$= 3\cos(2\sqrt{2}t) + 5\sqrt{2} \cdot \frac{1}{2\sqrt{2}}\sin(2\sqrt{2}t)$$

$$= 3\cos(2\sqrt{2}t) + \frac{5}{2}\sin(2\sqrt{2}t)$$

(2) $L^{-1}\left\{\dfrac{s+2}{s^2-4s+13}\right\}$

Solution: We have

$$L^{-1}\left\{\frac{s+2}{s^2-4s+13}\right\} = L^{-1}\left\{\frac{s-2+4}{(s-2)^2+3^2}\right\} = L^{-1}\left\{\frac{s-2}{(s-2)^2+3^2}\right\} + L^{-1}\left\{\frac{4}{(s-2)^2+3^2}\right\}$$

$$= e^{2t}\cos 3t + \frac{4}{3}e^{2t}\sin 3t = e^{2t}\left(\cos 3t + \frac{4}{3}\sin 3t\right)$$

(3) $L^{-1}\left\{\dfrac{s^4+s^2+1}{s^5}\right\}$

Solution:

$$L^{-1}\left\{\frac{1}{s}+\frac{1}{s^3}+\frac{1}{s^5}\right\} = L^{-1}\left\{\frac{1}{s}\right\} + \frac{1}{2!}L^{-1}\left\{\frac{2!}{s^3}\right\} + \frac{1}{4!}L^{-1}\left\{\frac{4!}{s^5}\right\}$$

$$= 1 + \frac{1}{2}t^2 + \frac{1}{24}t^4$$

(4) $L^{-1}\left\{\dfrac{s}{(s-2)^5}\right\}$

Solution: We have $L^{-1}\left\{\dfrac{s}{(s-2)^5}\right\} = L^{-1}\left\{\dfrac{s-2+2}{(s-2)^5}\right\}$

$$= L^{-1}\left\{\frac{1}{(s-2)^4} + \frac{2}{(s-2)^5}\right\}$$

$$= \frac{1}{3!}L^{-1}\left\{\frac{3!}{(s-2)^4}\right\} + \frac{2}{4!}L^{-1}\left\{\frac{4!}{(s-2)^5}\right\}$$

$$= \frac{1}{6}e^{2t}t^3 + \frac{1}{12}e^{2t}t^4$$

$$= \frac{1}{12}e^{2t}[2t^3+t^4]$$

(5) $L^{-1}\left\{\dfrac{s^2}{(s^2+a^2)^2}\right\}$

Solution: Consider

$$L^{-1}\left\{\dfrac{s^2}{(s^2+a^2)^2}\right\}=L^{-1}\left\{\dfrac{s^2+a^2-a^2}{(s^2+a^2)^2}\right\}$$

$$=L^{-1}\left\{\dfrac{1}{s^2+a^2}-\dfrac{a^2}{(s^2+a^2)^2}\right\}$$

$$=\dfrac{1}{a}L^{-1}\left\{\dfrac{a}{s^2-a^2}\right\}-\dfrac{1}{2a}L^{-1}\left\{\dfrac{2a^3}{(s^2+a^2)^2}\right\}$$

$$=\dfrac{1}{a}\sin at-\dfrac{1}{2a}(\sin at-at\cos at)$$

(6) $L^{-1}\left\{\dfrac{2s-1}{s^2-2s+10}\right\}$

Solution: Consider

$$L^{-1}\left\{\dfrac{2s-1}{s^2-2s+10}\right\}=L^{-1}\left\{\dfrac{2s-2+1}{(s-1)^2+3^2}\right\}$$

$$=2L^{-1}\left\{\dfrac{s-1}{(s-1)^2+3^2}\right\}+\dfrac{1}{3}L^{-1}\left\{\dfrac{3}{(s-1)^2+3^2}\right\}$$

$$=2e^t\cos 3t+\dfrac{1}{3}e^t\sin 3t$$

$$=\dfrac{2}{3}e^t(3\cos 3t+\sin 3t)$$

(7) Find the inverse Laplace transform of $\dfrac{2s^2-4}{(s+1)(s-2)(s-3)}$.

Solution: By partial fractions

Let $\dfrac{2s^2-4}{(s+1)(s-2)(s-3)}=\dfrac{A}{s+1}+\dfrac{B}{s-2}+\dfrac{C}{s-3}$

$$\therefore \ 2s^2 - 4 = A(s-2)(s-3) + B(s+1)(s-3) + C(s+1)(s-2)$$

$\therefore$ Put $s = -1$ we get $-2 = A(-3)(-4)$ $\therefore$ $A = -\dfrac{1}{6}$

Put $s = 2$ we get $4 = B(3)(-1)$ $\therefore$ $B = -\dfrac{4}{3}$

and put $s = 3$ we get $14 = C(4)(1)$ $C = \dfrac{7}{2}$

$$\therefore \quad f(t) = L^{-1}\left\{ \frac{2s^2 - 4}{(s+1)(s-2)(s-3)} \right\}$$

$$= -\frac{1}{6} L^{-1}\left\{ \frac{1}{s+1} \right\} - \frac{4}{3} L^{-1}\left\{ \frac{1}{s-2} \right\} + \frac{7}{2} L^{-1}\left\{ \frac{1}{s-3} \right\}$$

$$= -\frac{1}{6} e^{-t} - \frac{4}{3} e^{2t} + \frac{7}{2} e^{3t}$$

(8) Find the inverse Laplace transform of $\dfrac{2s^2 - 6s + 5}{s^3 - 6s^2 + 11s - 6}$

Solution: By partial fractions

Let $\dfrac{2s^2 - 6s + 5}{(s-1)(s-2)(s-3)} = \dfrac{A}{s-1} + \dfrac{B}{s-2} + \dfrac{C}{s-3}$

$$\therefore \quad 2s^2 - 6s + 5 = A(s-2)(s-3) + B(s-1)(s-3) + C(s-1)(s-2)$$

Put $s = 1$ $\qquad\qquad\qquad$ $1 = A(-1)(-2)$ $\qquad\qquad\qquad$ $\therefore A = \dfrac{1}{2}$

$s = 2$ $\qquad\qquad\qquad$ $1 = B(1)(-1)$ $\qquad\qquad\qquad$ $\therefore B = -1$

$s = 3$ $\qquad\qquad\qquad$ $5 = C(2)(1)$ $\qquad\qquad\qquad$ $C = \dfrac{5}{2}$

$$\therefore \quad f(t) = L^{-1}\left\{ \frac{2s^2 - 6s + 5}{(s-1)(s-2)(s-3)} \right\}$$

$$= \frac{1}{2} L^{-1}\left\{ \frac{1}{s-1} \right\} - 1 L^{-1}\left\{ \frac{1}{s-2} \right\} + \frac{5}{2} L^{-1}\left\{ \frac{1}{s-3} \right\}$$

$$= \frac{1}{2}e^t - e^{2t} + \frac{5}{2}e^{3t}$$

(9) Find the inverse L.T. of $\dfrac{2s+3}{(s-1)(s+2)^2}$.

Solution: By partial fractions

Let $\dfrac{2s+3}{(s-1)(s+2)^2} = \dfrac{A}{s-1} + \dfrac{B}{s+2} + \dfrac{C}{(s+2)^2}$

$\therefore \quad 2s+3 = A(s+2)^2 + B(s-1)(s+2) + C(s-1)$

Putting $s = 1, 2$, and 0 respectively, we get $A = \dfrac{5}{9}, B = -\dfrac{5}{9}, C = \dfrac{1}{3}$.

$$\therefore \qquad f(t) = \frac{5}{9}L^{-1}\left\{\frac{1}{s-1}\right\} - \frac{5}{9}L^{-1}\left\{\frac{1}{s+2}\right\} + \frac{1}{3}L^{-1}\left\{\frac{1}{(s+2)^2}\right\}$$

$$= \frac{5}{9}e^t - \frac{5}{9}e^{-2t} + \frac{1}{3}te^{-2t}.$$

(10) Find the inverse L.T. of $\dfrac{5s+3}{(s-1)(s^2+2s+5)}$.

Solution: By partial fractions

Let $\qquad \dfrac{5s+3}{(s-1)(s^2+2s+5)} = \dfrac{A}{s-1} + \dfrac{Bs+C}{s^2+2s+5}$

$\therefore \quad 5s+3 = A(s^2+2s+5) + (Bs+C)(s-1) = As^2 + 2As + 5A + Bs^2 + Cs - Bs - C$

$$= (A+B)s^2 + (2A-B+C)s + (5A-C)$$

Equating the coefficients

$$\left.\begin{array}{l} A+B=0 \\ 2A-B+C=5 \\ 5A-C=3 \end{array}\right\} \text{ Solving, } A = 1, B = -1 \text{ and } C = 2.$$

$$\therefore \quad f(t) = L^{-1}\left\{\frac{5s+3}{(s-1)(s^2+2s+5)}\right\} = L^{-1}\left\{\frac{1}{s-1}\right\} + L^{-1}\left\{\frac{-s+2}{(s+1)^2+2^2}\right\}$$

$$= L^{-1}\left\{\frac{1}{s-1}\right\} - L^{-1}\left\{\frac{s+1}{(s+1)^2+2^2}\right\} + 3L^{-1}\left\{\frac{1}{(s+1)^2+2^2}\right\}$$

$$= e^t - e^{-t}\cos 2t + \frac{3}{2}e^{-t}\sin 2t$$

(11) Find the inverse L.T. of $\dfrac{s^2+2}{(s^2+10)(s^2+20)}$

Solution:

Put $s^2 = p$ since the fraction involves only even powers of s.

By partial fractions

$$\therefore \quad \frac{p+2}{(p+10)(p+20)} = \frac{A}{p+10} + \frac{B}{p+20}$$

$$\therefore \quad p+2 = A(p+20) + B(p+10)$$

Put $p = -10$ $\qquad -8 = A(10)$ $\qquad \therefore \quad A = -\dfrac{4}{5}$

Put $p = -20$ $\qquad -18 = B(-10)$ $\quad \therefore \quad B = \dfrac{9}{5}$

$$\therefore \quad f(t) = -\frac{4}{5}L^{-1}\left\{\frac{1}{s^2+10}\right\} + \frac{9}{5}L^{-1}\left\{\frac{1}{s^2+20}\right\}$$

$$= \frac{-4}{5\sqrt{10}}L^{-1}\left\{\frac{\sqrt{10}}{s^2+(\sqrt{10})^2}\right\} + \frac{9}{5}\frac{1}{\sqrt{20}}L^{-1}\left\{\frac{\sqrt{20}}{s^2+(\sqrt{20})^2}\right\}$$

$$= \frac{-4}{5\sqrt{10}}\sin\sqrt{10}t + \frac{9}{5\sqrt{20}}\sin\sqrt{20}t$$

(12) Find the inverse L.T. of $\dfrac{s}{(s-3)(s^2+4)}$

Solution: By partial fraction

Let $\dfrac{s}{(s-3)(s^2+4)} = \dfrac{A}{s-3} + \dfrac{Bs+C}{s^2+4}$

Solving by *PFs*, we get

$$A = \frac{3}{13}, \; B = \frac{-3}{13}, \; C = \frac{4}{13}$$

$$\therefore \quad f(t) = \frac{3}{13} L^{-1}\left\{\frac{1}{s-3}\right\} - \frac{3}{13} L^{-1}\left\{\frac{s}{s^2+2^2}\right\} + \frac{2}{13} L^{-1}\left\{\frac{2}{s^2+2^2}\right\}$$

$$= \frac{1}{13}[3e^{3t} - 3\cos 2t + 2\sin 2t]$$

(13) Find the inverse Laplace transform of the following:

(i) $L^{-1}\left\{\log \dfrac{s+a}{s+b}\right\}$

Solution:

We use the formula $f(t) = -\dfrac{1}{t} L^{-1}\left\{\dfrac{d}{ds} F(s)\right\}$

$$\therefore \qquad f(t) = -\frac{1}{t} L^{-1}\left\{\frac{d}{ds}\left(\log \frac{s+a}{s+b}\right)\right\}$$

$$= -\frac{1}{t} L^{-1}\left\{\frac{d}{ds}(\log s + a - \log s + b)\right\}$$

$$= -\frac{1}{t} L^{-1}\left\{\frac{1}{s+a} - \frac{1}{s+b}\right\}$$

$$= -\frac{1}{t}(e^{-at} - e^{-bt}) = \frac{e^{-bt} - e^{-at}}{t}$$

(ii) $L^{-1}\left\{\dfrac{1}{2}\log \dfrac{s^2-a^2}{s^2+b^2}\right\}$

Solution: We have

$$f(t) = -\frac{1}{t} L^{-1}\left\{\frac{d}{ds}\frac{1}{2}\log \frac{s^2+a^2}{s^2+b^2}\right\}$$

$$= \frac{1}{t} L^{-1}\left\{\frac{1}{2}\frac{d}{ds}(\log s^2 + a^2 - \log s^2 + b^2)\right\}$$

$$= -\frac{1}{t} L^{-1} \frac{1}{2}\left(\frac{2s}{s^2+a^2} - \frac{2s}{s^2+b^2}\right)$$

$$= -\frac{1}{t} L^{-1}\left(\frac{s}{s^2+a^2} - \frac{s}{s^2+b^2}\right)$$

$$= -\frac{1}{t}(\cos at - \cos bt)$$

$$f(t) = \frac{\cos bt - \cos at}{t}$$

(iii) $L^{-1}\left\{\cot^{-1}\dfrac{s}{a}\right\}$

Solution: We have

$$f(t) = -\frac{1}{t} L^{-1}\left\{\frac{d}{ds}\cot^{-1}\frac{s}{a}\right\}$$

$$= -\frac{1}{t} L^{-1}\left\{\frac{-1}{1+\dfrac{s^2}{a^2}}\frac{1}{a}\right\}$$

$$= +\frac{1}{t} L^{-1}\left\{\frac{a}{a^2+s^2}\right\} = \frac{1}{t}\sin at$$

$$\therefore \quad f(t) = \frac{\sin at}{t}$$

(iv) $L^{-1}\left\{s\log\dfrac{s+1}{s-1}+2\right\}$

Solution: Consider

$$F(s) = s(\log s + 1 - \log s - 1) + 2$$

$$\therefore \qquad F'(s) = s\left(\frac{1}{s+1} - \frac{1}{s-1}\right) + \log\frac{s+1}{s-1}$$

$$= -\frac{2s}{s^2-1} + \log\frac{s+1}{s-1}$$

$$\therefore \qquad f(t) = -\frac{1}{t} L^{-1}\{F'(s)\}$$

$$= -\frac{1}{t}L^{-1}\left\{\frac{2s}{s^2-1}\right\} - \frac{1}{t}L^{-1}\left\{\log\frac{s+1}{s-1}\right\}$$

$$= \frac{2}{t}\cosh t - \frac{1}{t}L^{-1}\left\{\log\frac{s+1}{s-1}\right\}$$

$$= \frac{2}{t}\cosh t - \frac{1}{t}\left[-\frac{1}{t}L^{-1}\left\{\frac{1}{s+1} - \frac{1}{s-1}\right\}\right]$$

$$= \frac{2}{t}\cosh t + \frac{1}{t^2}(e^{-t} - e^{t})$$

$$= \frac{2}{t^2}(t\cosh t - \sinh t)$$

(v) $L^{-1}\left\{\dfrac{1}{s(s^2+a^2)}\right\}$

Solution: Here

$$F(s) = \frac{1}{s^2+a^2} \quad \therefore f(t) = \frac{1}{a}\sin at$$

$\therefore \qquad L^{-1}\left\{\dfrac{F(s)}{s}\right\} = \int\limits_0^t f(u)\,du$

$$= \frac{1}{a}\int\limits_0^t \sin at\; dt$$

$$= \frac{1}{a}\left(\frac{-\cos at}{a}\right)_0^a$$

$$= \frac{1}{a^2}(-\cos at + 1)$$

$$= \frac{1}{a^2}(1 - \cos at)$$

(vi) $L^{-1}\left\{\log\left(1 + \dfrac{4}{s^2}\right)\right\}$

Solution: By using the formula

$$f(t) = -\frac{1}{t} L^{-1} \left\{ \frac{d}{ds} \log \frac{s^2 + 4}{s^2} \right\}$$

$$= -\frac{1}{t} L^{-1} \left\{ \frac{d}{ds} \left(\log s^2 + 4 \right) - 2 \log s \right\}$$

$$= -\frac{1}{t} L^{-1} \left\{ \frac{2s}{s^2 + 4} - \frac{2}{s} \right\}$$

$$= \frac{-2}{t} (\cos 2t - 1)$$

$$\therefore \qquad f(t) = \frac{2(1 - \cos 2t)}{t}$$

(14) Find the inverse Laplace transform of

(a) $L^{-1} \left\{ \dfrac{e^{-2s}}{s - 3} \right\}$

Solution:

By second shifting theorem, we have

$$L^{-1}\{e^{-as} F(s)\} = f(t - a) H(t - a)$$

Here $\overline{F}(s) = \dfrac{1}{s - 3}$ then $f(t) = e^{3t}$

$$\therefore \qquad L^{-1} \left\{ \frac{e^{-2s}}{s - 3} \right\} = L^{-1}\{e^{-2s} F(s)\} = f(t - 2) H(t - 2)$$

$$= e^{3(t-2)} H(t - 2)$$

(b) $L^{-1} \left\{ \dfrac{s e^{-as}}{s^2 - w^2} \right\} \qquad a > 0$

Solution: Here

$$F(s) = \frac{s}{s^2 - w^2} \text{ then } f(t) = \cosh wt$$

$$\therefore \qquad L^{-1} \left\{ \frac{s e^{-as}}{s^2 - w^2} \right\} = L^{-1}\{e^{-as} F(s)\}$$

$$= f(t-a)H(t-a)$$

$$= \cosh w(t-a)H(t-a)$$

(c) $L^{-1}\left\{\dfrac{se^{-s/2} + \pi e^{-s}}{s^2 + \pi^2}\right\}$

Solution:

We have $L^{-1}\left\{e^{-s/2}\,\dfrac{s}{s^2 + \pi^2} + \dfrac{\pi e^{-s}}{s^2 + \pi^2}\right\}$

But $\qquad L^{-1}\left\{\dfrac{s}{s^2 + \pi^2}\right\} = \cos \pi t$ and $L^{-1}\left\{\dfrac{\pi}{s^2 + \pi^2}\right\} = \sin \pi t$

$$\therefore \qquad L^{-1}\left\{\frac{se^{-s/2} + \pi e^{-s}}{s^2 + \pi^2}\right\} = L^{-1}\left\{\frac{se^{-s/2}}{s^2 + \pi^2}\right\} + L^{-1}\left\{\frac{\pi e^{-s}}{s^2 + \pi^2}\right\}$$

$$= \cos \pi\left(t - \frac{1}{2}\right) H\left(t - \frac{1}{2}\right) + \sin \pi(t-1) H(t-1)$$

$$= \cos\left(\pi t - \frac{\pi}{2}\right) H\left(t - \frac{1}{2}\right) + \sin(\pi t - \pi) H(t-1)$$

$$= \sin \pi t\, H\left(t - \frac{1}{2}\right) - \sin \pi t\, H(t-1)$$

$$= \sin \pi t\left[H\left(t - \frac{1}{2}\right) - H(t-1)\right]$$

CONVOLUTION THEOREM

If $L^{-1}\{F(s)\} = f(t)$ and $L^{-1}\{G(s)\} = g(t)$, then

$$L^{-1}\{F(s)\cdot G(s)\} = \int_0^t f(u)g(t-u)\,du$$

Proof: Given $\qquad L^{-1}\{F(s)\} = f(t)$ and $L^{-1}\{G(s)\} = g(t)$

$$\therefore \quad F(s) = L\{f(t)\} = \int_0^\infty e^{-st} f(t)\,dt$$

and $G(s) = L\{g(t)\} = \int\limits_{0}^{\infty} e^{-st} g(t)dt$

To prove the theorem, it is sufficient to prove that

$$L\left\{\int\limits_{0}^{t} f(u)g(t-u)du\right\} = F(s) \cdot G(s)$$

Consider,

$$L\left\{\int\limits_{0}^{t} f(u)g(t-u)du\right\}$$

$$= \int\limits_{0}^{\infty} e^{-st}\left[\int\limits_{0}^{t} f(u)g(t-u)du\right]dt$$

$$= \int\limits_{t=0}^{\infty} \int\limits_{u=0}^{t} e^{-st} f(u)g(t-u)du\,dt$$

The domain of integration for double integral is from $u = 0$ to $u = t$ and $t = 0$ to $t = \infty$ as in Fig. (1). The double integration given in (1) indicates that we integrate first parallel to u-axis and then parallel to t-axis. We shall change the order of integration. Integrate first parallel to t-axis and then parallel to u-axis.

Parallel to t-axis the limits are $t = u$ to $t = \infty$ and parallel to u-axis are $u = 0$ to $u = \infty$.

$$= \int\limits_{0}^{\infty} f(u)\left[\int\limits_{u}^{\infty} e^{-st} g(t-u)dt\right]du$$

$$= \int\limits_{0}^{\infty} f(u)\,e^{-su}\left[e^{-(t-u)s} g(t-u)dt\right]du$$

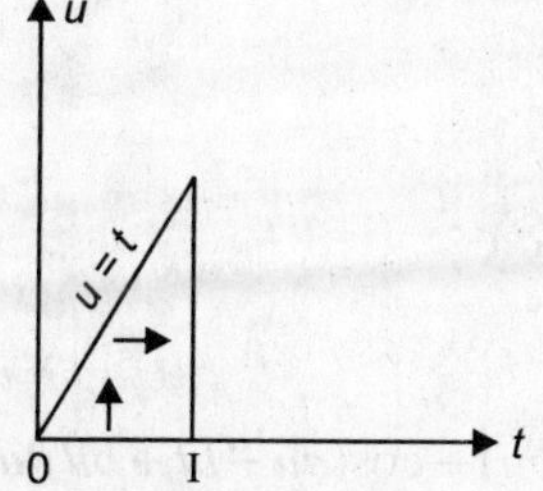

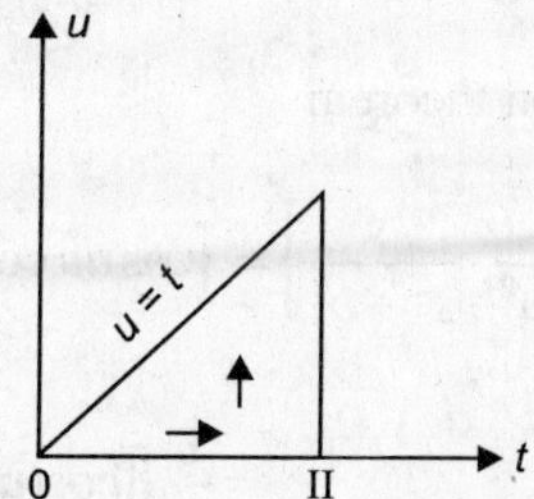

Put
$$\begin{array}{c|cc} t-u=v & t=u & v=0 \\ dt=dv & t=\infty & v=\infty \end{array}$$

$$= \int_0^\infty e^{-su} f(u) \left[\int_0^\infty e^{-sv} g(v)\,dv \right] du$$

$$= \int_0^\infty e^{-su} f(u)\,du \int_0^\infty e^{-sv} g(v)\,dv$$

$$= F(s)\cdot G(s)$$

$$\therefore \quad L^{-1}\{F(s)\cdot G(s)\} = \int_0^t f(u)g(t-u)\,du$$

Hence, the proof.

Examples

Evaluate the inverse Laplace transforms of the following by using convolution theorem.

(1) $L^{-1}\left\{\dfrac{s^2}{(s^2+a^2)(s^2+b^2)}\right\}$

Solution: Given function can be written as

$$L^{-1}\left\{\frac{s^2}{s^2+a^2}\cdot\frac{s}{s^2+b^2}\right\}$$

$$\therefore \quad f(t)=L^{-1}\left\{\frac{s}{s^2+a^2}\right\}=\cos at\Big|_{t\to u}$$

and $g(t)=L^{-1}\left\{\dfrac{s}{s^2+b^2}\right\}=\cos bt\Big|_{t\to(t-u)}$

By convolution theorem

$$L^{-1}\left\{\frac{s}{s^2+a^2}\cdot\frac{s}{s^2+b^2}\right\}=\int_0^t \cos au\,\cos b(t-u)\,du$$

$$=\frac{1}{2}\int_0^t [\cos(au+bt-bu)+\cos(au-bt+bu)]\,du$$

$$= \frac{1}{2} \int_0^t [\cos(a-b)u + bt) + \cos((a+b)u - bt)]$$

$$= \frac{1}{2} \left[\frac{\sin((a-b)u + bt)}{a-b} + \frac{\sin((a+b)u - bt)}{a+b} \right]_0^t$$

$$= \frac{1}{2} \left[\frac{\sin at}{a-b} + \frac{\sin at}{a+b} - \frac{\sin bt}{a-b} + \frac{\sin bt}{a+b} \right]$$

$$= \frac{1}{2} \left[\left(\frac{1}{a-b} + \frac{1}{a+b} \right) \sin at - \left(\frac{1}{a-b} - \frac{1}{a+b} \right) \sin bt \right]$$

$$= \frac{1}{2} \left[\frac{2a}{a^2 - b^2} \sin at - \frac{2b}{a^2 - b^2} \sin bt \right]$$

$$= \frac{1}{a^2 - b^2} [a \sin at - b \sin bt]$$

(2) $L^{-1} \left\{ \dfrac{s}{(s^2 + 2^2)(s^2 + 3^2)} \right\}$

Solution: Given function is

$$L^{-1} \left\{ \frac{s}{s^2 + 2^2} \cdot \frac{1}{s^2 + 3^2} \right\}$$

$$\therefore \qquad f(t) = L^{-1}\{F(s)\} = L^{-1} \left\{ \frac{s}{s^2 + 2^2} \right\} = \cos 2t \Big|_{t \to (t-4)}$$

and $\qquad g(t) = L^{-1}\{G(s)\} = L^{-1} \left\{ \dfrac{1}{s^2 + 3^2} \right\} = \dfrac{1}{3} \sin 3t \Big|_{t \to u}$

By convolution theorem

$$L^{-1} \left\{ \frac{s}{(s^2 + 2^2)(s^2 + 3^2)} \right\} = \frac{1}{3} \int_0^t \sin 3u \cos 2(t-u)\,du$$

$$= \frac{1}{3} \int_0^t \frac{1}{2} [\sin(3u + 2t - 2u) + \sin(3u - 2t + 2u)]\,du$$

$$= \frac{1}{6} \int_0^t [\sin(u + 2t) + \sin(5u - 2t)]\,du$$

$$= \frac{1}{6}\left[-\cos(u+2t) - \frac{\cos(5u-2t)}{5}\right]_0^t$$

$$= \frac{1}{6}\left[-\cos 3t - \frac{\cos 3t}{5} + \cos 2t + \frac{\cos 2t}{5}\right]$$

$$= \frac{1}{6}\left[\left(1+\frac{1}{5}\right)\cos 2t - \left(1+\frac{1}{5}\right)\cos 3t\right]$$

$$= \frac{1}{6}\left[\frac{6}{5}\cos 2t - \frac{6}{5}\cos 3t\right]$$

$$= \frac{1}{5}(\cos 2t - \cos 3t)$$

(3) $L^{-1}\left\{\dfrac{1}{s^2(s^2+16)}\right\}$

Solution: Given function can be written as

$$L^{-1}\left\{\frac{1}{s^2}\cdot\frac{1}{s^2+16}\right\}$$

$\therefore \qquad f(t) = L^{-1}\left\{\dfrac{1}{s^2}\right\} = t\Big|_{t\to(t-u)}$

$$g(t) = L^{-1}\left\{\frac{1}{s^2+4^2}\right\} = \frac{1}{4}\sin 4t\Big|_{t\to u}$$

By convolution theorem

$\therefore \qquad L^{-1}\left\{\dfrac{1}{s^2(s^2+16)}\right\}$

$$= \int_0^t \frac{1}{4}\sin 4u(t-u)\,du$$

$$= \frac{1}{4}\int_0^t (t-u)\sin 4u\,du$$

$$= \frac{1}{4}\left[(t-u)\left(\frac{-\cos 4u}{4}\right) - (-1)\frac{(-\sin 4u)}{4^2}\right]_0^t$$

$$= \frac{1}{4}\left[\frac{-\sin 4t}{16} + \frac{t}{4}\right]$$

$$= \frac{1}{64}[4t - \sin 4t]$$

(4) $L^{-1}\left\{\dfrac{s}{(s^2+a^2)^2}\right\}$

Solution: Given function is

$$L^{-1}\left\{\frac{1}{s^2+a^2} \cdot \frac{s}{s^2+a^2}\right\}$$

$$\therefore \quad f(t) = L^{-1}\left\{\frac{1}{s^2+a^2}\right\} = \frac{1}{a}\sin at \Big|_{t\to u}$$

and $g(t) = L^{-1}\left\{\dfrac{s}{s^2+a^2}\right\} = \cos at \Big|_{t\to(t-u)}$

By convolution theorem

$$\therefore \quad L^{-1}\left\{\frac{s}{(s^2+a^2)^2}\right\} = \frac{1}{a}\int_0^t \sin au \cos(at-au)\,du$$

$$= \frac{1}{2a}\int_0^t [\sin(au+at-au) + \sin(au-at+au)]\,du$$

$$= \frac{1}{2a}\int_0^t [\sin at + \sin(2au-at)]\,du$$

$$= \frac{1}{2a}\left[u\sin at - \frac{\cos(2au-at)}{2a}\right]_0^t$$

$$= \frac{1}{2a}\left[t\sin at - \frac{\cos(at)}{2a} + \frac{\cos at}{2a}\right]$$

$$= \frac{1}{2a}t\sin at$$

(5) $L^{-1}\left\{\dfrac{1}{s^2(s+2)^2}\right\}$

Solution: Given function is

$$L^{-1}\left\{\dfrac{1}{s^2}\cdot\dfrac{1}{(s+2)^2}\right\}$$

$$f(t)=L^{-1}\left\{\dfrac{1}{(s^2+2)^2}\right\}=e^{-2t}t\Big|_{t\to u}$$

and

$$g(t)=L^{-1}\left\{\dfrac{1}{s^2}\right\}=t\Big|_{t\to(t-u)}$$

By convolution theorem

$$\therefore\quad L^{-1}\left\{\dfrac{1}{s^2(s+2)^2}\right\}=\int_0^t e^{-2u}u(t-u)du$$

$$=\int_0^t(ut-u^2)e^{-2u}du$$

$$=\left[(ut-u^2)\dfrac{e^{-2u}}{-2}-(t-2u)\dfrac{e^{-2u}}{4}\right]_0^t$$

$$=\dfrac{te^{-2t}}{4}+\dfrac{t}{4}$$

$$=\dfrac{t}{4}(e^{-2t}+1)$$

(6) $L^{-1}\left\{\dfrac{1}{(s-4)(s+5)}\right\}$

Solution: Given function is $L^{-1}\left\{\dfrac{1}{(s-4)}\dfrac{1}{(s+5)}\right\}$

$$\therefore\quad f(t)=L^{-1}\left\{\dfrac{1}{s-4}\right\}=e^{4t}\Big|_{t\to u}$$

and
$$g(t) = L^{-1}\left\{\frac{1}{s+5}\right\} = e^{-5t}\Big|_{t\to(t-u)}$$

By convolution theorem

$$L^{-1}\left\{\frac{1}{(s-4)(s+5)}\right\} = \int_0^t e^{4u} \cdot e^{-5(t-u)}\, du$$

$$= \int_0^t e^{9u-5t}\, du$$

$$= \left[\frac{e^{9u-5t}}{9}\right]_0^t$$

$$= \frac{1}{9}[e^{ut} - e^{-5t}]$$

SOLUTION OF DIFFERENTIAL EQUATIONS USING LAPLACE TRANSFORMS

The Laplace transform method of solving differential equations gives particular solutions without the necessity of first finding the general solution and then evaluating the arbitrary constants. This method is, in general, shorter than our earlier methods and is specially useful for solving linear differential equations with constant coefficients.

Working Procedure

Working procedure to solve a linear differential equation with constant coefficients by transform method:

(1) Laplace transform has to be taken on both sides of the differential equation, using the formula of derivative and the given initial conditions.

(2) All the terms with negative sign are transposed to right.

(3) Divide by the coefficient of $L\{y(t)\}$ getting $L\{y(t)\}$ as a known function of s.

(4) Resolve this function of s into partial fractions.

(5) Take the inverse Laplace transform on both sides.

Then we get y as a function t which is the required solution satisfying the given conditions.

Note: We use the following formula of derivative in every example.

$$L\{y^n(t)\} = s^n L\{y(t)\} - s^{n-1}y(0) - s^{n-2}y'(0) \dots - y^{n-1}(0)$$

(i) $L\{y'(t)\} = sL\{y(t)\} - y(0)$

(ii) $L\{y''(t)\} = s^2 L\{y(t)\} - sy(0) - y'(0)$

(iii) $L\{y'''(t)\} = s^3 L\{y(t)\} - s^2 y(0) - sy'(0) - y''(0)$

Examples

(1) Solve the equation by the method of transform

$$\frac{d^2 y}{dt^2} - 2\frac{dy}{dt} + y = e^t \text{ given } y(0) = 2, \ y'(0) = -1.$$

Solution: The given differential equation is

$$y''(t) - 2y'(t) + y(t) = e^t$$

Taking Laplace transform on both sides

$$L\{y''(t)\} - 2L\{y'(t)\} + L\{y(t)\} = L\{e^t\}$$

$\therefore \quad s^2 L\{y(t) - sy(0) - y'(0) - 2[sL\{y(t)\} - y(0)] + L\{y(t)\} = \dfrac{1}{s-1}$

$\therefore \quad (s^2 - 2s + 1)L\{y(t)\} = \dfrac{1}{s-1} + 2s - 5$

$\therefore \quad L\{y(t)\} = \dfrac{1}{(s-1)^3} + \dfrac{2s-5}{(s-1)^2}$

$$= \frac{1}{(s-1)^3} + \frac{2(s-1)}{(s-1)^2} - \frac{3}{(s-1)^2}$$

$$= \frac{1}{(s-1)^3} + \frac{2}{s-1} - \frac{3}{(s-1)^2}$$

$\therefore \quad y(t) = L^{-1}\left\{\dfrac{1}{(s-1)^3}\right\} + 2L^{-1}\left\{\dfrac{1}{s-1}\right\} - 3L^{-1}\left\{\dfrac{1}{(s-1)^2}\right\}$

$$= \frac{1}{2}e^t t^2 + 2e^t - 3e^t t$$

$\therefore \quad y = e^t \left(\dfrac{t^2}{2} - 3t + 2\right)$

(2) Solve $y'' + 9y = 18t$ given $y(0) = 0$ and $y(\pi/2) = 0$ by Laplace transform method.

Solution: Given differential equation is

$$y''(t) + 9y(t) = 18t$$

Taking Laplace transform on both sides

$$L\{y''(t)\} + 9L\{y(t)\} = 18L\{t\}$$

$$\therefore \quad s^2 L\{y(t)\} - sy(0) - y'(0) + 9L\{y(t)\} = 18\frac{1}{s^2}$$

Take $y'(0) = k$ and $y(0) = 0$

$$\therefore \qquad (s^2 + 9)L\{y(t)\} = \frac{18}{s^2} + k$$

$$\therefore \qquad L\{y(t)\} = \frac{18}{s^2(s^2 + 9)} + \frac{k}{s^2 + 9}$$

$$\therefore \qquad y(t) = 18L^{-1}\left\{\frac{1}{s^2(s^2 + 9)}\right\} + kL^{-1}\left\{\frac{1}{s^2 + 9}\right\}$$

Let $\dfrac{1}{s^2(s^2 + 9)} = \dfrac{A}{s} + \dfrac{B}{s^2} + \dfrac{Cs + D}{s^2 + 9}$, by partial fraction

$$1 = As(s^2 + 9) + B(s^2 + 9) + (Cs + D)s^2$$

$$1 = (A + C)s^3 + (B + D)s^2 + 9As + 9B$$

Equating the coefficients

$$A + C = 0 \quad \rightarrow \quad C = 0$$

$$B + D = 0 \quad \rightarrow \quad D = -\frac{1}{9}$$

$$9A = 0 \quad \rightarrow \quad A = 0$$

$$9B = 1 \quad \rightarrow \quad B = \frac{1}{9}$$

$$\therefore \quad y(t) = 18 \cdot \frac{1}{9} L^{-1}\left\{\frac{1}{s^2}\right\} - 18 \cdot \frac{1}{9} L^{-1}\left\{\frac{1}{s^2 + 9}\right\} + kL^{-1}\left\{\frac{1}{s^2 + 9}\right\}$$

$$y(t) = 2t - \frac{2}{3}\sin 3t + \frac{k}{3}\sin 3t$$

when $t = \dfrac{\pi}{2}$, $y\left(\dfrac{\pi}{2}\right) = 0$

$$\therefore \qquad 0 = 2 \cdot \dfrac{\pi}{2} - \dfrac{2}{3}\sin\dfrac{3\pi}{2} + \dfrac{k}{3}\sin\dfrac{3\pi}{2}$$

$$\therefore \qquad k = 3\pi + 2$$

$$\therefore \qquad y = 2t - \dfrac{2}{3}\sin 3t + \dfrac{(3\pi + 2)}{3}\sin 3t$$

$$y = 2t + \pi\sin 3t \ \text{ is the required solution.}$$

(3) Solve by Laplace transform method

$$\dfrac{d^2 y}{dt^2} + w^2 y = \cos wt \ \ t > 0 \ \text{ given } \ y(0) = y'(0) = 0$$

Solution: The given differential equation is

$$y''(t) + w^2 y(t) = \cos wt$$

Taking Laplace transform on both sides

$$L\{y''(t)\} + w^2 L\{y(t)\} = L\{\cos wt\}$$

$$s^2 L\{y(t)\} - sy(0) - y'(0) + w^2 \ L\{y(t)\} = \dfrac{s}{s^2 + w^2}$$

$\therefore$ Given $y(0) = y'(0) = 0$

$$\therefore \quad (s^2 + w^2)L\{y(t)\} = \dfrac{s}{s^2 + w^2}$$

$$\therefore \quad y(t) = L^{-1}\left\{\dfrac{s}{(s^2 + w^2)^2}\right\}$$

$$= \dfrac{1}{2w} L^{-1}\left\{\dfrac{2ws}{(s^2 + w^2)^2}\right\}$$

$$\therefore \qquad y = \dfrac{1}{2w} t \sin wt$$

(4) Solve $\dfrac{d^3 y}{dx^3} + 2\dfrac{d^2 y}{dx^2} - \dfrac{dy}{dx} - 2y = 0$ given $y = 1$, $\dfrac{dy}{dx} = 2$, $\dfrac{d^2 y}{dx^2} = 2$ at $x = 0$ by Laplace transform method.

Solution: Given equation is

$$y'''(x) + 2y''(x) - y'(x) - 2y(x) = 0$$

Taking Laplace transform on both sides

$$[s^3 L\{y(x)\} - s^2 y(0) - sy'(0) - y''(0)] + 2[s^2 L\{y(x)\} - sy(0) - y'(0)]$$

$$-[sL\{y(x)\} - y(0)] - 2L\{y(x)\} = 0$$

$$\therefore \quad (s^3 + 2s^2 - s - 2)L\{y(x)\} = s^2 + 2s + 2 + 2s + 4 - 1$$

$$= s^2 + 4s + 5$$

$$\therefore \quad L\{y(x)\} = \frac{s^2 + 4s + 5}{(s+2)(s^2 - 1)}$$

$$= \frac{s^2 + 4s + 5}{(s+2)(s+1)(s-1)}$$

$$\therefore \quad y(x) = L^{-1}\left\{\frac{s^2 + 4s + 5}{(s+2)(s+1)(s-1)}\right\}$$

Let $\dfrac{s^2 + 4s + 5}{(s+2)(s+1)(s-1)} = \dfrac{A}{s+2} + \dfrac{B}{s+1} + \dfrac{C}{s-1}$, by partial fraction

$$\therefore \quad s^2 + 4s + 5 = A(s+1)(s-1) + B(s+2)(s-1) + c(s+2)(s+1)$$

put

$$s = -2 \quad 1 = 3A \quad A = \frac{1}{3}$$

$$s = -1 \quad 2 = 2B \quad B = -1$$

$$s = 1 \quad 10 = 6C \quad C = \frac{5}{3}$$

$$\therefore \quad y(x) = \frac{1}{3}L^{-1}\left\{\frac{1}{s+2}\right\} - 1L^{-1}\left\{\frac{1}{s+1}\right\} + \frac{5}{3}L^{-1}\left\{\frac{1}{s-1}\right\}$$

$$y = \frac{1}{3}e^{-2x} - e^{-x} + \frac{5}{3}e^{x}$$

(5) Solve $\dfrac{d^2 x}{dt^2} + 2\dfrac{dx}{dt} + 5x = e^{-t}\sin t$ given $x(0) = 0$, $x'(0) = 1$ by Laplace transform method.

Solution: Given equation is

$$x''(t) + 2x'(t) + 5x(t) = e^{-t}\sin t$$

Taking Laplace transform on both sides

$$L\{x''(t)\} + 2L\{x'(t)\} + 5L\{x(t)\} = L\{e^{-t}\sin t\}$$

$$\therefore \quad S^2 L\{x(t)\} - s(x(0) - x'(0) + 2[sL\{x(t)\} - x(0)] + 5L\{x(t)\} = \frac{1}{(s+1)^2 + 1^2}$$

$$\therefore \quad (s^2 + 2s + 5)L(x(t)\} = \frac{1}{(s^2 + 2s + 2)} + 1$$

$$\therefore \quad L\{x(t)\} = \frac{1}{(s^2 + 2s + 5)(s^2 + 2s + 2)} + \frac{1}{s^2 + 2s + 5}$$

$$\therefore \quad x(t) = L^{-1}\left\{\frac{1}{(s^2 + 2s + 5)(s^2 + 2s + 2)}\right\} + L^{-1}\left\{\frac{1}{s^2 + 2s + 5}\right\}$$

$$= \frac{1}{3}L^{-1}\left\{\frac{1}{s^2 + 2s + 2} - \frac{1}{s^2 + 2s + 5}\right\} + L^{-1}\left\{\frac{1}{s^2 + 2s + 5}\right\}$$

$$= \frac{1}{3}L\left\{\frac{1}{s^2 + 2s + 2}\right\} + \frac{2}{3}L^{-1}\left\{\frac{1}{s^2 + 2s + 5}\right\}$$

$$= \frac{1}{3}L\left\{\frac{1}{(s+1)^2 + 1^2}\right\} + \frac{1}{3}L^{-1}\left\{\frac{2}{(s+1)^2 + 2^2}\right\}$$

$$= \frac{1}{3}e^{-t}(\sin t + \sin 2t)$$

(6) Solve $y'' + 6y' + 9y = 12t^2 e^{-3t}$ given $y(0) = y'(0) = 0$ by Laplace transform method.

Solution: Given equation is

$$y''(t) + 6y'(t) + 9y(t) = 12t^2 e^{-3t}$$

Taking Laplace transform on both sides

$$L\{y''(t)\} + 6L\{y'(t)\} + 9L\{y(t)\} = 12L\{t^2 e^{-3t}\}$$

$$\therefore \quad s^2 L\{y(t)\} - sy(0) - y'(0) + 6[sL\{y(t)\} - y(0)] + 9L\{y(t)\} = 12 \cdot \frac{2}{(s+3)^3}$$

$$\therefore \quad (s^2 + 6s + 9)L\{y(t)\} = \frac{24}{(s+3)^3}$$

$$\therefore \quad (s+3)^2 L\{y(t)\} = \frac{24}{(s+3)^3}$$

$$\therefore \quad y(t) = 24L^{-1}\left\{\frac{1}{(s+3)^5}\right\}$$

$$= L^{-1}\frac{4!}{(s+3)^5} = e^{-3t}t^4$$

$$\therefore \quad y = t^4 e^{-3t}$$

SIMULTANEOUS DIFFERENTIAL EQUATIONS

Engineering experiments involving two or more dependent variables and only one independent variable results in simultaneous differential equations which can be solved by applying Laplace transform.

Examples

(1) Solve the simultaneous equations by Laplace transform method

$$\frac{dx}{dt} + y = \sin t; \quad \frac{dy}{dt} + x = \cos t$$

given that $x(0) = 2$, $y(0) = 0$.

Solution: The given equations are

$$x'(t) + y(t) = \sin t$$

$$y'(t) + x(t) = \cos t$$

Taking Laplace transform on both sides

$$sL\{x(t)\} - x(0) + L\{y(t)\} = \frac{1}{s^2+1}$$

$$sL\{y(t)\} - y(0) + L\{x(t)\} = \frac{s}{s^2+1}$$

$$\therefore \quad sL(x(t)\} + L\{y(t)\} = \frac{1}{s^2+1} + 2 \qquad (1)$$

$$sL(y(t)\} + Lx(t)\} = \frac{s}{s^2+1} \tag{2}$$

Multiply (1) by s and then subtract from (2)

$$(s^2-1)L\{x(t)\} = 2s$$

$$\therefore \qquad L\{x(t)\} = \frac{2s}{s^2-1}$$

$$\therefore \qquad x = 2L^{-1}\left\{\frac{s}{s^2-1}\right\} = 2\cosh t$$

and

$$y = \sin t - \frac{d}{dt}(2\cosh t)$$

$$= \sin t - 2\sinh t$$

$$\therefore \qquad x = 2\cosh t, \ y = \sin t - 2\sinh t$$

(2) Solve $\dfrac{dx}{dt} + \dfrac{dy}{dt} + x = -e^{-t}$

$$\frac{dx}{dt} + 2\frac{dy}{dt} + 2x + 2y = 0 \text{ subject to the conditions}$$

$$x(0) = -1, \ y(0) = 1 \text{ by L.T. method}$$

Solution: Given equations are

$$x'(t) + y'(t) + x(t) = -e^{-t}$$

$$x'(t) + 2y'(t) + 2x(t) + 2y(t) = 0$$

Taking Laplace transform to both the equations

$$sL\{x(t)\} - x(0) + sL\{y(t)\} - y(0) + L\{x(t)\} = -\frac{1}{s+1}$$

$$sL\{x(t)\} - x(0) + 2sL\{y(t)\} - 2y(0) + 2L\{x(t)\} + 2L\{y(t)\} = 0$$

$$\therefore \qquad (s+1)L\{x(t)\} + sL\{y(t)\} = \frac{-1}{s+1} \tag{1}$$

$$(s+2)L\{x(t)\} + 2(s+1)L\{y(t)\} = 1 \tag{2}$$

Multiply (1) by $2(s + 1)$ and (2) by s

$$\therefore \qquad 2(s+1)^2 L\{x(t)\} + 2s(s+1)L\{y(t)\} = -2$$

$$(s+2)sL\{x(t)\} + 2s(s+1)L\{y(t)\} = s$$

Subtracting,

$$[2(s+1)^2 - s(s+2)]L\{x(t)\} = -s-2$$

$$\therefore \qquad (s^2 + 2s + 2)L\{x(t)\} = -(s+2)$$

$$\therefore \qquad L\{x(t)\} = -\frac{(s+2)}{s^2 + 2s + 2}$$

$$\therefore \qquad x(t) = -L^{-1}\left\{\frac{s+1}{(s+1)^2 + 1^2} + \frac{1}{(s+1)^2 + 1^2}\right\}$$

$$= -e^{-t}(\cos t + \sin t)$$

and $y(t) = e^{-t}(1 + \sin t)$.

(3) Solve $\dfrac{d^2 x}{dt^2} + 9x = \cos 2t$ given $x(0) = 1$, $x(\pi/2) = -1$ by Laplace transform method.

Solution: Given equation is

$$x''(t) + 9x(t) = \cos 2t$$

Taking L.T. on both sides

$$Lx''(t) + 9L\{x(t)\} = L\{\cos 2t\}$$

$$s^2 L\{x(t)\} - 5x(0) - x'(0) + 9L\{x(t)\} = \frac{s}{s^2 + 2^2}$$

Take $x'(0) = k$

$$\therefore \qquad (s^2 + 9)L\{x(t)\} = s + k + \frac{s}{s^2 + 4}$$

$$\therefore \qquad L\{x(t)\} = \frac{s+k}{s^2 + 9} + \frac{s}{(s^2 + 4)(s^2 + 9)} = \frac{As+B}{s^2 + 4} + \frac{Cs+D}{s^2 + 9}$$

$$\therefore \qquad x(t) = L^{-1}\left\{\frac{s}{s^2 + 9}\right\} + kL^{-1}\left\{\frac{1}{s^2 + 9}\right\} + \frac{1}{5}L^{-1}\left(\frac{s}{s^2 + 4}\right) - \frac{1}{5}L^{-1}\left\{\frac{s}{s^2 + 9}\right\}$$

by P.Fs.

$$= kL^{-1}\left\{\frac{1}{s^2+9}\right\} + \frac{1}{5}L^{-1}\left\{\frac{s}{s^2+4}\right\} - \frac{4}{5}L^{-1}\left\{\frac{s}{s^2+9}\right\}$$

$$x = \frac{k}{3}\sin 3t + \frac{1}{5}\cos 2t + \frac{4}{5}\cos 3t$$

When $t = \pi/2$ $x(\pi/2) = -1$

$$-1 = -\frac{k}{3} - \frac{1}{5} \text{ or } \frac{k}{3} = \frac{4}{5}$$

$\therefore$ $x = \dfrac{1}{5}(\cos 2t + 4\sin 3t + 4\cos 3t)$ is the required solution.

APPLICATIONS

Examples

(1) A volt Ee^{-at} is applied at $t = 0$ to a circuit of inductance L and resistance R, show by L.T. method the current at time is

$$\frac{E}{R-aL}\left[e^{-at} - e^{-\frac{kt}{L}}\right] \text{ and } i(0) = 0$$

Solution: The differential equation

$$L\frac{di}{dt} + Ri(t) = Ee^{-at}$$

$\therefore$ $$Li'(t) + Ri(t) = Ee^{-at}$$

Taking L.T. on both sides

$$L[sL\{i(t)\} - i(0)] + R\{i(t)\} = EL\{e^{-at}\}$$

$\therefore$ $$(Ls + R)L\{i(t)\} = \frac{E}{s+a}$$

$\therefore$ $$i(t) = L^{-1}\left\{\frac{E}{(s+a)(Ls+R)}\right\}$$

$$= EL^{-1}\left\{\frac{A}{s+a} + \frac{B}{Ls+R}\right\}$$

$$\therefore \quad A = \frac{1}{R-aL} \quad B = \frac{L}{aL-R} = \frac{-L}{R-aL}$$

$$\therefore \quad i(t) = \frac{E}{R-aL} \; L^{-1}\left\{\frac{1}{s+a} - \frac{1}{s+\dfrac{R}{L}}\right\}$$

$$\therefore \quad i = \frac{E}{R-aL}\left(e^{-at} - e^{-\frac{Rt}{L}}\right)$$

(2) The differential equation of a uniformly loaded beam with simple supports is

$$EI\frac{d^2 y}{dx^2} = -\frac{wx}{2}(l-x)$$

where EI and w are constants. Solve by L.T. method given that $y(0) = y'(0) = 0$.

Solution: The given equation is

$$y''(x) = -\frac{w}{2EI}(lx - x^2)$$

Taking L.T. on both sides

$$s^2 L\{y(x)\} - sy(0) - y'(0) = -\frac{w}{2EI}\left(\frac{l}{s^2} - \frac{2}{s^3}\right)$$

$$\therefore \quad L\{y(x)\} = -\frac{w}{2EI}\left(\frac{l}{s^4} - \frac{2}{s^5}\right)$$

$$\therefore \quad y(x) = -\frac{w}{2EI}L^{-1}\left\{\frac{1}{s^4} - \frac{2}{s^5}\right\}$$

$$= -\frac{w}{2EI}\left(\frac{l}{3!}x^3 - \frac{2}{4!}x^4\right)$$

$$\therefore \quad y = -\frac{w}{2EI}\left(\frac{x^3}{6} - \frac{x^4}{12}\right)$$

(3) Spring can extend 20 cm when 0.5 kg mass is attached to it. It is suspended vertically from a support and set into vibration by putting it down 10 cm and imposing a velocity 5 cm/sec vertically upwards. Find the displacement from its equilibrium position at time t.

Solution: In the equilibrium position, we have $mg = kb$ where b is the extension. When $m = 0.5$ kg $= 500$ gram, $b = 20$ cm, taking $g = 980$ cm/sec^2, we have

$$k = \frac{mg}{b} = \frac{500 \times 980}{20} = 24500 \text{ dynes/cm}$$

Equation of motion is

$$\frac{d^2 x}{dt^2} + \frac{k}{m} x = 0$$

i.e.

$$\frac{d^2 x}{dt^2} + \frac{24500}{500} x = 0$$

$\therefore$

$$\frac{d^2 x}{dt^2} + 49x = 0$$

Taking L.T. on both sides

$$s^2 L\{x(t)\} - sx(0) - x'(0) + 49L\{x(t)\} = 0$$

Initially, the mass was at a depth of 10 cm below the equilibrium position $\therefore$ $x(0) = 10$.

Also the initial velocity was 5 cm/sec. vertically upwards $\therefore$ $x'(0) = -5$

$\therefore$

$$(s^2 + 49)L(x(t)) = 105 - 5$$

$\therefore$

$$Lx(t)\} = \frac{105 - 5}{s^2 + 7^2}$$

$\therefore$

$$x(t) = 10L^{-1}\left\{\frac{s}{s^2 + 7}\right\} - 5L^{-1}\left\{\frac{1}{s^2 + 7^2}\right\}$$

$$= 10\cos 7t - \frac{5}{7}\sin 7t$$

$\therefore$

$$x = 10\cos 7t - \frac{5}{7}\sin 7t$$

QUESTION BANK 6

(A) Evaluate the inverse Laplace transform of

(1) $L^{-1}\left\{\log\frac{1+s}{s}\right\}$ $\qquad$ Ans: $\frac{1}{t}(1 - e^{-t})$

(2) $L^{-1}\left\{\frac{1}{s(s^2 + 4)}\right\}$ $\qquad$ Ans: $\frac{1}{4}(1 - \cos 2t)$

(3) $L^{-1}\left\{\dfrac{s^2}{s^4-a^4}\right\}$ 　　　　　Ans: $\dfrac{1}{2a}(\sinh at+\sin at)$

(4) $L^{-1}\left\{\cot^{-1}(s+1)\right\}$ 　　　　　Ans: $e^{-t}\dfrac{\sin t}{t}$

(5) $L^{-1}\left\{\log\dfrac{s(s+1)}{s^2+4}\right\}$ 　　　　　Ans: $\dfrac{2\cos 2t-e^{-t}-1}{t}$

(6) $L^{-1}\left\{\dfrac{1}{s(s^2-a^2)}\right\}$ 　　　　　Ans: $\dfrac{1}{a^2}(\cosh at-1)$

(7) $L^{-1}\left\{\dfrac{e^{-\pi s}}{s^2+1}\right\}$ 　　　　　Ans: $-\sin t\,H(t-\pi)$

(8) $L^{-1}\left\{\dfrac{e^{-\pi s}}{s^2}\right\}$ 　　　　　Ans: $(t-\pi)H(t-\pi)$

(9) $L^{-1}\left\{\dfrac{e^{-ks}}{s^2(s+a)}\right\}$ 　　　　　Ans: $\dfrac{1}{a^2}[a(t-k)+1+e^{-a(t-k)}]H(t-k)$

(10) $L^{-1}\left\{\dfrac{s+2}{s^2(s+1)(s-2)}\right\}$ 　　　　　Ans: $\dfrac{1}{3}(e^{2t}-e^{-t}-t)$

(11) Solve the following equations by the transform method.

(i)　　$y''-3y'+2y=4t+3e^{3t}$ when $y(0)=1$ and $y'(0)=-1$

$$\text{Ans: } y=2t+3+\frac{1}{2}(e^{3t}-e^{t})-2e^{2t}$$

(ii)　　$y''+2y'-3y=\sin t$ given $y=0,\ y'=0$ when $t=0$

$$\text{Ans: } y=\frac{1}{8}e^{t}-\frac{1}{40}e^{-3t}-\frac{1}{10}(2\sin t+\cos t)$$

(iii)　　$y''+y=t$ given $y(0)=1,\ y'(0)=-2$

$$\text{Ans: } y=t-3\sin t+\cos t$$

(iv)　　$\dfrac{dx}{dt}-y=e^{t},\ \dfrac{dy}{dt}+x=\sin t$ given $x(0)=1,\ y(0)=0$

$$\text{Ans: } y=\frac{1}{2}(t\sin t-e^{t}+\cos t-\sin t)$$

$$x=\frac{1}{2}(e^{t}+\cos t+2\sin t-t\cos t)$$

(v) The current i_1 and i_2 in mesh are given by the equations

$$\frac{di_1}{dt} - wi_2 = a\cos pt; \quad \frac{di_2}{dt} - wi_1 = a\sin pt; \text{ find } i_1 \text{ and } i_2 \text{ if } i_1 = i_2 = 0$$

when $t = 0$.

$$\text{Ans: } i_1 = \frac{a}{p+w}(\sin wt + \sin pt), \ i_2 = \frac{a}{p+w}(\cos wt - \cos pt)$$

(B) (1) If $L\{f(t)\} = F(s)$ then prove that $L\left\{\dfrac{f(t)}{t}\right\} = \displaystyle\int_s^\infty F(s)\,ds$

(2) Find (i) $L^{-1}\left\{\dfrac{s+3}{s^2 - 6s + 13}\right\}$ (ii) $L^{-1}\left\{\dfrac{4s+5}{(s-1)^2(s+2)}\right\}$

  (iii) $L^{-1}\left\{\dfrac{s+1}{s-1}\right\}$

(3) Find (i) $L^{-1}\left\{\dfrac{s^2 - 3s + 1}{s^3}\right\}$ (ii) $L^{-1}\left\{\dfrac{5s-3}{(s-1)(s^2 + 2s + 5)}\right\}$

  (iii) $L^{-1}\left\{\log\dfrac{s+1}{s}\right\}$

(4) Find (i) $L^{-1}\left\{\log\dfrac{s+a}{s+b}\right\}$ (ii) $L^{-1}\left\{\dfrac{s+2}{s^2 + 4s + 5}\right\}$

  (iii) $L^{-1}\left\{\dfrac{s^2 + 10s + 13}{(s-1)(s^2 - 5s + 6)}\right\}$

(5) Find (i) $L^{-1}\left\{\dfrac{2s+1}{(s+2)(s-1)^2}\right\}$ (ii) $L^{-1}\left\{\dfrac{1}{(s^2 + 4)^2}\right\}$

  (iii) $L^{-1}\left\{\log\left(1 + \dfrac{a^2}{s^2}\right)\right\}$

(6) If $L\{f(t)\} = F(s)$ then prove that

$$L\{f''(t)\} = s^2 L\{f(t) - sf(0) - f''(0)$$

(7) If $L\{f(t)\} = F(s)$ then prove that

$$L\{f'(t)\} = sF(s) - f(0)$$

(8) Using Laplace transform method, solve

$$\frac{d^2y}{dt^2} + 4\frac{dy}{dt} + 3y = e^{-t} \text{ with } y(0) = y'(0) = 1$$

(9) Solve the following equation using Laplace transform method

$$y'' - 3y' + 2y = 12e^{-t} \text{ given that } y(0) = 2,\ y'(0) = 6$$

(10) Using the Laplace transform technique, solve

$$\frac{d^2x}{dt^2} - 2\frac{dx}{dt} + x = e^t,\ x(0) = 2 \text{ and } \frac{dx}{dt} = -1 \text{ at } x = 0$$

(11) Solve $\dfrac{d^2y}{dt^2} + 4\dfrac{dy}{dt} + 4y = e^{-t}$, $y(0) = 0$, $y'(0) = 0$ by Laplace transform method.

(12) Using Laplace transform, solve

$$\frac{d^2y}{dt^2} - 2\frac{dy}{dt} + y = e^{-t} \text{ given that } y = 2,\ \frac{dy}{dt} = -1 \text{ at } t = 0$$

(13) Solve the differential equation

$$y'' + 2y' + y = 6te^{-t} \text{ under the condition } y(0) = 0 = y'(0)$$

using Laplace transform.

(14) Solve by Laplace transform method

$$\frac{d^2y}{dt^2} + w^2 y = \cos wt \ \ t > 0 \text{ given } y(0) = y'(0) = 0$$

(15) Solve the simultaneous equations using Laplace transforms

$$\frac{dx}{dt} + y = \sin t,\ \frac{dy}{dt} + x = \cos t \text{ given that } x = 2,\ y = 0 \text{ for } t = 0 .$$

(16) Using Laplace transform method, solve, the following simultaneous equations.

$$3\frac{dx}{dt} + \frac{dy}{dt} + 2x = 1$$

$$\frac{dx}{dt} + 4\frac{dy}{dt} + 3y = 0 \text{ given } x = 0,\ y = 0 \text{ at } t = 0$$

(17) Solve the simultaneous differential equations by Laplace transform method

$$\frac{dx}{dt} + 5x - 2y = t$$

$$\frac{dy}{dt} + 2x + y = 0 \text{ given that } x = y = 0 \text{ when } t = 0$$

Fourth Semester B.E. Degree Examination, May/June 2010

Advanced Mathematics-II

Time: 3 hrs.

Max. Marks: 100

Note: Answer any Five full questions.

1. a. Find the projection of the line AB on CD where
   $A = (1, 2, 3)$, $B = (-1, 0, 2)$, $C = (1, 4, 2)$, $D = (2, 0, -1)$. (06 marks)
   b. Find the angle between two lines whose direction cosines
   are given by $l + 3\,m + 5\,n = 0$ and $2\,mn - 6\,nl - 5\,lm = 0$ (07 Marks)
   c. A line makes angles α, β, γ, δ with diagonals of a cube.

   Prove that $\cos^2\alpha + \cos^2\beta + \cos^2\gamma + \cos^2\delta = \dfrac{4}{3}$ (07 Marks)

2. a. Find the equation of the plane passing through the points
   $(3, 1, 2)$ and $(3, 4, 4)$ and perpendicular to $5x + y + 4z = 0$. (06 Marks)
   b. Show that the points $(2, 2, 0)$, $(4, 5, 1)$, $(3, 9, 4)$ and
   $(0, -1, -1)$ are coplanar. Find the equation of the plane
   containing them. (07 Marks)
   c. Find the equation of a straight line through $(7, 2, -3)$ and
   perpendicular to each of the lines.

   $$\frac{x-2}{3} = \frac{y-3}{4} = \frac{z-4}{5} \text{ and } \frac{x+2}{4} = \frac{y-3}{5} = \frac{z-4}{6}$$ (07 Marks)

3. a. Show that the position vectors of the vertices of a triangle

   $\vec{a} = 3(\sqrt{3}\hat{i} - \hat{j})$, $\vec{b} = 6\hat{j}$, $\vec{c} = 3(\sqrt{3}\hat{i} + \hat{j})$ form an isosceles
   triangle. (06 Marks)
   b. A particle moves along the curve

   $\vec{r} = 3t^2\hat{i} + (t^3 - 4t)\hat{j} + (3t + 4)\hat{k}$. Find the components

   of velocity and acceleration at $t = 2$ in the direction

   $\hat{i} - 2\hat{j} + 2\hat{k}$. (07 Marks)
   c. Find the angle between the normals to the surfaces
   $x^2y^2 = z^4$ at $(1, 1, 1)$ and $(3, 3, -3)$. (07 Marks)

4. a. Find the directional derivatives of the function $\phi = xyz$
   along the direction of the normal to the surface
   $xy^2 + yz^2 + zx^2 = 3$ at the point $(1, 1, 1)$. (06 Marks)

b. Find the div $\vec{F}$ and curl $\vec{F}$ where

$$\vec{F} = \nabla(x^3 + y^3 + z^3 - 3xyz).$$ (07 Marks)

c. If $\vec{v} = 2xy\,\hat{i} + 3x^2 y\,\hat{j} - 3ayz\,\hat{k}$ is solenoidal at (1, 1, 1),
   find a.   (07 Marks)

5. a. Find the unit normal vector to the surface
   $xy + x + zx = 3$ at (1, 1, 1).   (06 Marks)

b. Find the constants 'a', 'b', 'c' such that the vector field

$$(\sin y + az)\hat{i} + (bx\cos y + z)\hat{j} + (x + cy)\hat{k}$$

is irrotational. Also find the scalar field ϕ such that $\vec{F} = \nabla\phi$.   (07 Marks)

c. Prove that $\nabla^2 (\log r) = \dfrac{1}{r^2}$

   where $\vec{r} = x\hat{i} + y\hat{j} + z\hat{k}$ and $r = |\vec{r}|$.   (07 Marks)

6. a. Find the Laplace transform of $\sin 2t \sin 3t$.   (05 Marks)

b. Find $L\left[\dfrac{(1 - e^t)}{t}\right]$.   (05 Marks)

c. Find $L[e^{-t}(3\sinh 2t - 2\cosh 3t)]$ (05 Marks)

d. Find the Laplace transform of

$$f(t) = \begin{cases} t/\lambda & \text{when} \quad 0 < t < \lambda \\ 1 & \text{when} \quad t > \lambda \end{cases}$$ (05 Marks)

7. a. Evaluate $\displaystyle\int_0^\infty \dfrac{\sin t}{t}\, dt$ using Laplace transform.   (05 Marks)

b. Find the inverse Laplace tranform of $\dfrac{1}{(s^2 + 3s + 2)(s + 3)}$.   (05 Marks)

c. Find $L^{-1}\left[\dfrac{s - 1}{s^2 - 6s + 25}\right]$.   (05 Marks)

d. Find $L^{-1}\left[\log\left\{\dfrac{s^2 + 1}{s^2 - s}\right\}\right]$.   (05 Marks)

8. a. Find $L^{-1}\left[\dfrac{1}{s^2(s + 5)}\right]$ using convolution theorem.   (10 Marks)

b. Solve the differential equation $y'' + 2y' + y = 6te^{-t}$ under
   the condition $y(0) = (0) = y'(0)$ using Laplace transform.   (10 Marks)

Fourth Semester B.E. Degree Examination

Advanced Mathematics-II

Time: 3 hrs.   Max. Marks: 100

Note: Answer any Five full questions.

1. a. Define direction cosines. Find the angle between two lines whose direction cosines are (l_1, m_1, n_1) and (l_2, m_2, n_2).

    b. Find the co-ordinates of the foot of the perpendicular from $A\,(1, 1, 1)$ to the line joining $B\,(1, 4, 6)$ and $C\,(5, 4, 4)$.

    c. If a line makes angles α, β, γ and δ with the four diagonals of a cube, show that

    $$\cos^2\alpha + \cos^2\beta + \cos^2\gamma + \cos^2\delta = \frac{4}{3}$$

    $(6 + 7 + 7 = 20$ Marks)

2. a. With usual notations, derive the equation of the plane

    $$\frac{x}{a} + \frac{y}{b} + \frac{z}{c} = 1$$

    b. Find the equation of the plane through the line of intersection of the planes $x + y + z = 1$ and $2x + 3y + 4z = 5$ and perpendicular to the plane $x - y + z = 0$

    c. Show that the lines $\dfrac{x-5}{4} = \dfrac{y-7}{4} = \dfrac{z+3}{-5}$ and

    $\dfrac{x-8}{7} = \dfrac{y-4}{1} = \dfrac{z-5}{3}$ intersect. Find their point of intersection and the equation of the plane in which they lie. $(6 + 7 + 7 = 20$ Marks)

3. a. Find a unit vector normal to both the vectors

    $4\hat{i} - \hat{j} + 3\hat{k}$ and $-2\hat{i} + \hat{j} - 2\hat{k}$. Find also the sine of the angle between them.

    b. Prove that $\vec{a}\times(\vec{b}\times\vec{c}) = (\vec{a}\cdot\vec{c})\vec{b} - (\vec{a}\cdot\vec{b})\vec{c}$.

    c. If $\vec{a}, \vec{b}, \vec{c}$ are any three vectors, prove that

    $$[\bar{b}\times\bar{c},\ \bar{c}\times\bar{a},\ \bar{a}\times\bar{b}] = [\bar{a}, \bar{b}, \bar{c}]^2$$

    $(6 + 7 + 7 = 20$ Marks)

4. a. A particle moves along a curve whose parameteric equations are
$x = e^{-t}$, $y = 2\cos 3t$, $z = 2\sin 3t$, where t is the time. Find the velocity and acceleration at any time t. Also find the initial velocity and initial accelerations.

b. Prove that $\nabla^2 r^n = n(n+1)r^{n-2}$.

c. Find div $\vec{F}$ and curl $\vec{F}$ if
$$\vec{F} = (3x^2 - 3yz)\hat{i} + (3y^2 - 3xz)\hat{j} + (3z^2 - 3xy)\hat{k} \quad (6 + 7 + 7 = 20 \text{ Marks})$$

5. a. Prove that $\nabla \times (\nabla \times \vec{F}) = \nabla(\nabla \cdot \vec{F}) - \nabla^2 \vec{F}$.

b. Find the angle between the surfaces $x^2 + y^2 + z^2 = 9$ and $x^2 + y^2 - z = 3$ at the point $(2, -1, 2)$.

c. Find the directional derivative of $\phi = xy^2 + yz^3$ at the point $(2, -1, 1)$ in the direction of $\hat{i} + 2\hat{j} + 2\hat{k}$. $(6 + 7 + 7 = 20 \text{ Marks})$

6. a. Find the Laplace transform of $f(t) = \sin at$.

b. Find (i) $L\{\sin^3 2t\}$ (ii) $L\left\{\dfrac{1-\cos at}{t}\right\}$

c. Find (i) $L\{t\sin at\}$ (ii) $L\{e^t \cosh 2t\}$ $(6 + 7 + 7 = 20 \text{ Marks})$

7. a. If $L\{f(t)\} = F(s)$ then prove that $L\left\{\displaystyle\int_0^t f(u)du\right\} = \dfrac{1}{s}F(s)$

b. Find (i) $L^{-1}\left\{\dfrac{s+3}{s^2-4s+13}\right\}$ (ii) $L^{-1}\left\{\dfrac{s^3+2s+5}{s^5}\right\}$

c. Find (i) $L^{-1}\left\{\dfrac{4s+5}{(s-1)^2(s+2)}\right\}$ (ii) $L^{-1}\left\{\log\dfrac{s+a}{s-b}\right\}$ $(6 + 7 + 7 = 20 \text{ Marks})$

8. a. Using Laplace transform method, solve
$$\dfrac{d^2y}{dt^2} + 4\dfrac{dy}{dt} + 3y = e^{-t} \text{ with } y(0) = y'(0) = 1$$

b. Solve the following simultaneous differential equations by Laplace transform method.
$$\dfrac{dx}{dt} + y = \sin t; \quad \dfrac{dy}{dt} + x = \cos t$$
Given $x = 1$, $y = 0$ when $t = 0$. $(10 + 10 = 20 \text{ Marks})$

Index